Computer-Driven Instructional Design with INTUITEL
An Intelligent Tutoring Interface for Technology-Enhanced Learning

RIVER PUBLISHERS SERIES IN INNOVATION AND CHANGE IN EDUCATION - CROSS-CULTURAL PERSPECTIVE

Volume 15

Indexing: All books published in this series are submitted to Thomson Reuters Book Citation Index (BkCI), CrossRef and to Google Scholar.

Nowadays, educational institutions are being challenged as professional competences and expertise become progressively more complex. This is mainly because problems are more technology-bounded, unstable and ill-defined with the involvement of various integrated issues. Solving these problems requires interdisciplinary knowledge, collaboration skills, and innovative thinking, among other competences. In order to facilitate students with the competences expected in their future professions, educational institutions worldwide are implementing innovations and changes in many respects.

This book series includes a list of research projects that document innovation and change in education. The topics range from organizational change, curriculum design and innovation, and pedagogy development to the role of teaching staff in the change process, students' performance in the areas of not only academic scores, but also learning processes and skills development such as problem solving creativity, communication, and quality issues, among others. An inter- or cross-cultural perspective is studied in this book series that includes three layers. First, research contexts in these books include different countries/regions with various educational traditions, systems and societal backgrounds in a global context. Second, the impact of professional and institutional cultures such as language, engineering, medicine and health, and teachers' education are also taken into consideration in these research projects. The third layer incorporates individual beliefs, perceptions, identity development and skills development in the learning processes, and inter-personal interaction and communication within the cultural contexts in the first two layers.

We strongly encourage you as an expert within this field to contribute with your research and help create an international awareness of this scientific subject.

For a list of other books in this series, visit www.riverpublishers.com

Computer-Driven Instructional Design with INTUITEL
An Intelligent Tutoring Interface for Technology-Enhanced Learning

Editors

Kevin Fuchs

University of Applied Sciences Karlsruhe
Germany

Peter A. Henning

University of Applied Sciences Karlsruhe
Germany

River Publishers

Routledge
Taylor & Francis Group
LONDON AND NEW YORK

Published 2017 by River Publishers

River Publishers
Alsbjergvej 10, 9260 Gistrup, Denmark
www.riverpublishers.com

Distributed exclusively by Routledge

4 Park Square, Milton Park, Abingdon, Oxon OX14 4RN
605 Third Avenue, New York, NY 10158

First published in paperback 2024

Computer-Driven Instructional Design with INTUITEL: An Intelligent Tutoring Interface for Technology-Enhanced Learning / by Kevin Fuchs, Peter A. Henning.

Routledge is an imprint of the Taylor & Francis Group, an informa business

ISBN: 978-87-93519-51-0 (hbk)
ISBN: 978-87-7004-434-9 (pbk)
ISBN: 978-1-003-33769-0 (ebk)

DOI: 10.1201/9781003337690

Contents

Preface

INTUITEL is a resarch project that was co-financed by the European Commission. It was conducted from 2012 to 2015 under the participation of 12 European partners: Karlsruhe University of Applied Sciences, Steinbeis-Europa-Zentrum, University of Reading, University of Vienna, International University of La Rioja, Holon Institute of Technology, Fraunhofer IOSB, TIE Kinetix, FZI Research Center for Information Technology, Universidad de Valladolid, Lattanzio Learning and IMC AG.

INTUITEL aims to advance state-of-the-art e-learning systems via addition of guidance and feedback for learners. Through a combination of pedagogical knowledge, measured learning progress and a broad range of environmental and background data, INTUITEL provides guidance towards an optimal learning pathway.

INTUITEL blends in with several free and commercial Learning Management Systems. The number of supported platforms is extendable via a clearly defined interface description that can be implemented by any Learning Management System.

This book is both a summary of the findings of the INTUITEL project and a guidance for developers who want to implement their own INTUITEL-enabled system.

List of Contributors

Alessandro Barberi, *University of Vienna, Austria*

Alexander Schmoelz, *University of Vienna, Austria*

Alexander Streicher, *Fraunhofer Institute for Optronics System Technology and Image Exploitation IOSB, Germany*

Alexandra Forstner, *University of Vienna, Austria*

Christian Swertz, *University of Vienna, Austria*

Daniel Burgos, *Universidad Internacional de La Rioja, Spain*

Elena Verdú, *Universidad Internacional de La Rioja, Spain*

Elisabetta Parodi, *Lattanzio Learning S.p.A., Italy*

Florian Heberle, *University of Applied Sciences Karlsruhe, Germany*

Juan P. de Castro, *Universidad de Valladolid, Spain*

Kevin Fuchs, *University of Applied Sciences Karlsruhe, Germany*

Luis de-la-Fuente-Valentin, *Universidad Internacional de La Rioja, Spain*

Luisa M. Regueras, *Universidad de Valladolid, Spain*

María J. Verdú, *Universidad de Valladolid, Spain*

Peter A. Henning, *University of Applied Sciences Karlsruhe, Germany*

Stefan Zander, *FZI Research Center for Information Technology, Germany*

Sven Steudter, *IMC AG, Germany*

Uta Schwertel, *IMC AG, Germany*

List of Figures

List of Tables

List of Abbreviations

CC	Concept Container
CCM	Cognitive Content Model
CL	Communication Layer
CM	Cognitive Model
DF	Didactic Factor
DM	Domain Model
GUI	Graphical User Interface
KD	Knowledge Domain
KO	Knowledge Object
KT	Knowledge Type
LMS	Learning Management System
LO	Learning Object
LORE	Learning Object Recommender
LP	Learning Pathway
LPM	Learner Progress Model
LPM	Learning Progress Model
MCS	Multidemensional Cognitive Space
mId	message identifier
MLP	Macro Learning Pathway
mLP	micro Learning Pathway
MT	Media Type
OWL	Web Ontology Language
PO	Pedagogical Ontology
QB	Query Builder
REST	Representational State Transfer
RR	Recommendation Rewriter
SCORM	Sharable Content Object Reference Model
SLOM	Semantic Learning Object Model
SMW	Semantic MediaWiki
TUG	Tutorial Guidance
USE	User Score Extraction
UUID	Universal Unique Identifier

1

Introduction

Kevin Fuchs

The imagination of machines replacing human teachers – maybe not as a whole but at least part-wise – has inspired philosophers educational and computer scientists for many decades. One of the first steps was Skinner's and Pressey's teaching machines – at that time purely mechanical instructional devices [78, 79, 92, 93]. Since then the desire accrued to create machines that are capable of empathizing with human beings. Empathy is what empowers a teacher to adopt the perspective of a student. Perspective adoption is the key ability that we need to explain things and to create individual instructions and recommendations.

In the case of human beings, Jean Piaget gave us a detailed description of the development of this particular skill. He observed children aged between three and 6 years developing the ability of adopting another person's perspective. From him, we know that the prerequisite that constitutes this ability is what we call self-consciousness, which means being conscious of oneself in contrast to others and the environment. We can adopt another one's perspective only if we are aware of our own point of view and the difference to the other ones.

In the field of learning and teaching we have contented ourselves with a concept that appears to be less ambitious than the one we call "consciousness". We rather use the term "adaptivity" to describe a machine's capability of adapting itself to the learner's individual needs. How can we create adaptive machines and algorithms, respectively? Indeed, there is an elegant way to give a universal answer. Alan Turing has bequeathed us the Turing machine – until today the strongest model of a universal state-based machine. If we can make a statement about the Turing machine this is also valid for any kind of machine we know – no matter what technology it is based on.

Turing's universal machine receives input from outside and transfers it into a particular internal state. Moreover, the Turing machine can output information that it may read again as its input. This way, the Turing machine may feed itself with changes of its internal state. In other words: The Turing machine has the capacity of self-back coupling. Indeed, this self-back coupling is the only mechanism we can facilitate to create adaptivity in machines. Consequently, if we intend to create adaptive learning environments, they have to be designed as elaborated in the following. The inner state of the system must comprise a knowledge representation of both the learning content and the student. Driven by outer events – for example certain actions and behaviors of the student – the knowledge representation is modified which equates to the transition to another internal state.

The system feeds itself now with a representation of this new internal state, inferring on it and generating the respective instructions and recommendations for the learner. This process is repeated cyclically with each modification of the knowledge base.

The INTUITEL research project – an acronym for "Intelligent Tutoring Interface for technology Enhanced Learning" aimed to develop a general design for such an adaptive system. In the first place, INTUITEL is a design pattern. In the second place, the research team also developed a prototype system to prove the functionality of the design. The aforementioned knowledge representation is based on ontologies. With these ontologies pedagogical and didactic knowledge of both the learning content and the learner is modeled. Outer events are triggered by the learner's actions and the periphery of the system. These events form the input that results in a modification of the inner state – more precisely the ontologies modeling the knowledge about the learner and the learning content. A reasoning unit acts as the transfer unit by inferring on the just modified ontologies and generating according learning recommendations for the learner. This adaption is performed cyclically with every change of the knowledge base.

These are the fundamental principles of the INTUITEL concept which we will discuss in this book. Not only will the reader become familiar with the theoretical foundations. We also give an instruction on how to build your own INTUITEL system.

2

Intelligent Tutoring Systems: Preliminary Thoughts

This chapter introduces the educational background of the INTUITEL concept. It argues from general considerations to actual problems. First, basic structural elements of organized teaching and learning processes and an interpretation of these elements for teaching and learning are laid out. Second, implications of using computer technology as a medium in organized teaching and learning processes are discussed. Third, the history of adaptive assistant systems for educational processes is presented. Fourth, conclusions from these preliminary thoughts are drawn.

2.1 Organized Teaching and Learning Processes

Christian Swertz, Alexander Schmoelz, Alessandro Barberi and Alexandra Forstner

Dead people don't learn. While this might read a bit too existentialistic for the beginning of a chapter about adaptive assistant systems for educational processes, it is helpful to open up two perspectives: First, computers do not live. Thus, they cannot learn. Second, learning is closely connected to being alive. While we will discuss the first point later on, the second one allows us to make some basic distinctions here.

In some theories, all processes are understood as an exchange of information, no matter if the processes take place in the context of a living being or in the context of some dead matter. If it's all just about an exchange of information, there is no clear criterion to distinguish between living beings and dead matter. The exchange of information in these theories just means that they are transferred from one process to another.

But the transfer of information can be understood in two ways: First, the transfer can be understood as copying. In that case, the receiver just

3

adds the received information to the information stored in the receiver. Second, transferring can be understood as understanding. In this understanding, information needs to be expressed as signs. Since the relation between signs and objects is arbitrary, signs need to be understood. In that case, the receiver interprets the information by adding meaning to it and the meaning is arbitrary.

Both types of transferring information are sometimes considered as learning. But they are hardly comparable and very different in nature. Thus, it is necessary to distinguish between both types of learning. Unfortunately in our context, the application of computer technology in teaching and learning, both types of transferring information are relevant. This sometimes seems to create a tendency to neglect the difference between both types of information transfer. Usually, the first type is addressed as machine learning and the second type is addressed as human learning. While the word learning occurs in both cases, it does not mean the same term in both cases. Both occurrences are sometimes treated as synonyms, but in fact they are homonyms, since dead people cannot learn and computer technology is dead matter. While transferring information takes place with computer technology, computer technology cannot learn in the sense of human learning.

Thus, our first distinction here is the distinction between transferring information as a copy process that does not require meaning making and human learning that requires the understanding of signs. For clearness' sake we will use learning only in the sense of human learning. What we mean by learning in the following chapters is the communication of knowledge among living beings, and considerably among human beings.

For human beings, learning is an existentialistic problem indeed. It is necessary to learn. Without learning, human beings cannot live. In this context, the subject of didactic is the organized teaching and learning of human beings. While there are quite some and quite different theories about teaching and learning, there is no doubt that humans beings need to be educated. And there is no doubt that they can be creative as well, despite the fact that there are different theories of creativity as well. These two premises, the need to be educated and creativity as basic qualities, are axioms since they cannot be doubted in educational research. Objecting these premises would mean to reject the possibility of education overall.

Starting with these premises and our basic distinction, we are exposing a set of theorems to explicate the educational perspective that was used while developing the INTUITEL approach in this chapter. These theorems are:

1. The future of human beings is open.
2. Human beings can learn to determine themselves.
3. Education is non-deterministic.
4. Education takes place among human beings (generations).
5. Education takes place in a community.
6. Learning processes cannot be observed.

With these theorems, we suggest a theoretical framework to design Adaptive Assistant Systems.

2.1.1 The Open Future of Human Beings

Education is necessary for human beings. But it is not possible to finally predict the results of education [88, 21]. One of the reasons is that human beings are always able to stand up against external influences. They have an own free will. This free will is something that cannot be turned off or overridden. It can be shaped by the context human beings live in, it can be influenced by social interests, other people can try to break it or get it under their control, and we can try to get rid of it, by using drugs, becoming religious fanatics or whatever. Attempts to get rid of the own free will seem to take place if the own free will becomes a burden, which might be the case if human beings are treated as dead matter. But in that case, education is kind of pointless anyway. And the attempt is useless, since it's an expression of the own free will. While we're alive and awake, the own free will is present at all times.

In education, the idea of a free will is often connected to the philosophy of Kant. While details of the relevance of a free will for education are disputed occasionally [33], a free will can't be finally doubted, since doubting a free will already postulates a free will. As a free will is a necessary presumption for education, the cause-effect relations are not suitable to understand education [46].

Since we have to assume a free will at least in some respect, the result of educational processes can't be finally predicated. It is possible to set up educational institutions, curricula, assessment systems, assistant systems, and stuff like that. And it is very much possible to get people to act as if they do not have an own free will in those contexts. But that's it. It's not possible to force people to really judge external influences as meaningful, no matter which motivational strategies, outcome definitions or whatever is brought into educational processes. On the other hand, people sometimes judge content or actions as relevant, even if they are not meant to be relevant. They agree, accept content or actions as relevant, and give it a meaning by making it part of their personality. But due to the own free will, it's not possible to finally predict that acceptance will happen. The only thing we might predict is that people will act as if they accepted it.

Thus, we are facing a fundamental tension among the necessity to be educated and the unpredictable results of educational processes here. This tension indicates the open future of educational processes. The tension between force and freedom is the starting point for our design of an adaptive assistant system.

Besides being open, educational processes are focused on the individual. At least since Comenius has published his didactic in 1657 with "omnes, omnia, omnino" on the title, the individual (and not the average) is important for education. This is necessarily the case, since learning can't be substituted. It's not possible to learn for somebody else. We need to learn ourselves. If somebody else learns something, we do not know anything and as already stated, we cannot just copy the learning results. We need to understand things ourselves. Of course, we can and have to rely on other people's understandings while learning. But to do so, we need to understand that we took the decision to rely on other people's understanding. And obviously, we have to learn how to take a decision like that, and this can't be substituted, since the decision can't be taken by somebody else. If somebody else takes the decision, somebody else will rely on other people's understanding and start to learn. So we can't get out of learning we have to learn and to decide for ourselves. Making meaning is inevitable.

The open future and the focus on the individual make it difficult to prove one teaching method as the very best and the only one for organized teaching and learning. To prove one teaching method as the very best one, it would be necessary to state that it is successful not only in the present, but will be successful in the future. While the prediction of future reactions might be possible with some likelihood in the case of stocks and stones, this is not the case for the behavior of an individual human being. People might decide to act in a predictable way for a while, but sometimes they suddenly change their mind and start to do something different and that's not happening in every century or so. Instead, it can be considered as taking place in anything we do, since we are hardly able to repeat an action exactly as we acted before.

Repeating an action exactly as it was done before might be imagined in a context where only logical operations exist. In that case, we are not facing actions, but something like carrying out orders. To be precise here, not even obeying orders is possible in a context of logical operations, since obeying orders require the possibility to reject orders. But a computer cannot say: "Hey folks, I'm sick and tired of opcode EA, I'm not doing it any more". A computer does not obey instructions. The instructions are just carried out, and they are repeated again and again if required. That's something human beings can't do, even if they want to. And sometimes, they decide to try something completely different instead.

That's why repeatability is hardly ever used as a criterion for scientific truth in the social sciences. We thus have to assume here, that the only certainty, if it comes to teaching and learning methods, is a negative one: It is not possible to prove one teaching method as the very best and only one. Still, teaching and learning methods are possible and might help to get people to act as if they like one teaching and learning method. Teaching and learning methods might also be accepted by human beings, at least temporarily, or in the long run. The criterion for an acceptance in the long run is that people use those teaching and learning methods themselves, that is: The methods are passed among generations. In this respect, educational theories fall into their subject area themselves.

Teaching methods are most often tied to certain research methods. While research and teaching methods are often presented as connected closely (like in [85] or [103], this is always problematic, since no research method can be proven as the eternally right one. The same applies to ethics. In turn, no teaching method, like programmed instruction [30], open learning [90] or pragmatic learning [52] can be proven as the only or best one by scientific research.

If teaching methods and research methods are tied together, teaching methods are connected to certain scientific paradigms [59]. The programmed instruction for example is tied to the paradigmatic experiments conducted by Skinner with pigeons while some theories of learning styles are connected to the Tilting Tests conducted by Witkin [31].

Theories of learning styles illustrate the problem that is at stake here. To show the problem, we have to consider the theories of learning styles as a phenomenon itself. The basic idea of learning style theories is that it is possible to determine people's learning style, present learning material according to the learning style and increase the learning outcome that way. Learning style theories thus assume that learning styles are relatively stable personality traits and that it is possible to predict future learning behavior, since it is necessary to conclude from a standardization procedure in the past to a learning process in the future.

Interestingly, this does not work. There is hardly any evidence for the idea that considering the results of a learning style inventory while designing teaching and learning processes improves learning outcomes [49, 50]. We can suggest three possible explanations for the failure of learning style theories here: First, learning is not influenced by personality traits only. There is a subject area with a certain structure that requires recognition, there are teachers, there is an administration, an institution, a family and so on. All these aspects are considered (consciously or not) by the learner. In other terms: the context matters. This does not mean that context theories are the very best solution. They are just another paradigm.

Second, learners learn how to learn while learning. If we assume that people learn and that learning styles are learned, it does not seem far-fetched to assume that people learn how to learn while they learn considering all that context, their personality and most probably also stuff they will never talk about, however deep you dig into the unconscious. If people learn to learn while they learn, they can do it all the time and thus change their learning style every now and then and it looks like they do it at moments we can't predict.

This might be connected to a third possible explanation. This can be coined as learning to the test. Teaching to the test as a strategy of teachers to prepare students for a standardized test is usually not highly esteemed. But learning to the test is something students do whatever the test looks like. If the test requires some ticks at the right answer, students will prepare for that. And if the test requires some holistic or even critical thinking, students will at least act as if they could think holistically or critically.

It's an interesting argument that they might at least somehow think critically in the second case, no matter what they do: If they accept the requirements of the test, they think critically. And if they only act as if they accepted the requirements, but rejected them instead, they think critically too. In any case, if there is a test, learners usually learn to the test. By doing so, they avoid challenging the procedures and rules of pedagogical institutions, and that's probably a good idea. In turn, this gives some further evidence for the thesis that it is not possible to predict the learning behavior of human beings. Thus, learning styles are no longer termed as personality traits, but as the expression of an individual's analysis of the learning environment by decision making agents.

While the limits and problems of learning style theories have been shown and argued quite often, learning style theories and learning style inventories are still pretty popular. Thus, they are a phenomenon that asks for explanation. But we are not going to suggest a theory of the tempting character of learning style theories or the personality traits of their followers here. Our point is that it is not possible to predict the learning behavior of human beings. Thus, it is also not possible to predict that a certain teaching method will create improved learning outcomes. For the design of teaching and learning environments, playing with multiple teaching and learning methods is more promising than mechanical reactions to test the results.

2.1.2 Learning to Determine Oneself

Play is a cultural phenomenon that appeared all through history. In ancient times, playing games had been considered as not very relevant. It appeared in paintings sometimes, but it is not emphasized as a relevant subject for theoretical discussions. In medieval times, playing games was considered as bad, since it degrades working power and promotes sin and vice [75]. An important change in the perception of games is expressed in Bruegels painting "Kinderspiele" (children's games), which was first shown in 1553. Playing games became considered more as a sphere with a value of its own. The right of people to play was accepted as long as playing contributes to something useful, like the stimulation of mental abilities [75].

This understanding of playing games was picked up in pedagogical considerations by Basedow in the eighteenth century [74]. Basedow suggested to convert all the games children play into something useful. Therefore, Basedow applied games to teach subjects like Latin or Biology. This idea to apply games for teaching something useful is still widespread today, particularly in concepts for digital game based learning [76] or serious games.

At the end of the eighteenth century, the understanding of games was changed and extended substantially. This change culminates in the famous words of Schiller [87]: "Denn, um es endlich auf einmal herauszusagen, der Mensch spielt nur, wo er in voller Bedeutung des Wortes Mensch ist, und er ist nur da ganz Mensch, wo er spielt [For, to finally speak it out at once, man only plays when he is a man in the full meaning of the word, and he is only completely man when he plays]". With this sentence, Schiller identified play as the area where people can become people, and thus as the central place for human development and education.

Schiller discussed this place in the context of arts. He considered arts as a context where human activities have to be understood as play. A necessary condition for this context is freedom, not usefulness. For Schiller, this freedom means being free of being forced by other people's reasoning (kings, priests, etc.) and of being forced by nature (food, housing, etc.). Being free from external forces opens up a room for creative actions, and these creative actions are by no means intended to be useful or profitable.

In our context, the important point is, that play as an existential aspect of human development fundamentally refers to human freedom and the own free will. Due to this, play cannot be controlled from the outside, but only be done by people themselves. This changes the pedagogical perspective in contrast to Basedow. Basedow tried to control learning processes by creating games. With Schiller, playing is understood as an activity that cannot be controlled. Still, playing needs some sort of playground. A room where playing is actually possible is needed, but it cannot be forced that a room for playing games is actually used to play. With Schiller's theory it is possible to understand teaching and learning as a game where people play with content – and where people play with media that are used to learn the content.

2.1.3 Education as a Non-Deterministic Process

Since freedom and the necessity for self-determination are essential parts of education, it's not possible to predict the results of teaching and learning in individual cases. And it's not possible to predict, which teaching activities are appropriate in which situation. Thus, teaching can't be guided by theory only. This problem was introduced by Herbart in 1802 into educational sciences. Herbart differentiates pedagogy into an academic discipline and an artistic practice. Academic theories are derived from principles and made of broad concepts. Artistic practice has to deal with individual circumstances.

While active educational artists (like teachers) like to refer to personal experiences and observations to justify their educational actions, this is – according to Herbart – nothing else than casualness (Schlendrian). Instead, a well-founded theory has to be used to guide observations and experiments. Additionally, Herbart states that studying an educational theory is helpful for guiding the art of education performed by actual teachers. Still, teachers need to act as teachers to actually learn how to be a teacher. In other words: being a teacher cannot be learned from theory alone, but is essentially connected to sharing a common social and, according to Herbart, artistic practice.

This idea of being a teacher is understood by Herbart with the concept of pedagogical attitudes (pädagogischer Takt). Even if the pedagogically acting artist is a profound theoretician, he is not able to consider all his theoretical knowledge while teaching, since he has to act immediately in actual situations. This time pressure makes it necessary to act intuitively while performing pedagogical artwork. Still, these pedagogical attitudes are not considered as everlasting attributes of the personality by Herbart, but as habits that can be changed by theoretical considerations as well as by different experiences. Thus, changing the intuition that is used by teachers is the central objective of teacher training programs for Herbart.

One of the consequences of this concept is, as Herbart points out, that educational actions cannot fully meet the requirements of each individual case. Thus, educational actions always fail at least partly. The possibility to fail is, therefore, a necessary aspect of performing educational actions. While Herbart was convinced that a complete theory of teaching and learning is possible (but not available to him), this conviction is no longer accepted in the educational sciences today. The principle of plurality [80] leads to the conclusion that there is more than one way of teaching and learning in any context.

From this point of view, the debate between behavioristic, constructivistic, instructionalistic or situated learning theories appears rather pointless, since learning actually takes place whichever approach is chosen. The relevant problem is rather to creatively combine objectives, content, methods, and media in a learning environment in meaningful ways. The act of combining objectives, content, methods, and media is understood as theory-practice transformation by Herbart. The theory of the theory-practice transformation indicates a dialectic between thinking and acting. This dialectic problem needs to be considered when designing learning environments with algorithms and data.

Important for us is Herbart's conclusion that the creation of meaningful environments requires intuitive actions, which are based on pedagogical attitudes and guided by pedagogical theories. We suggest understanding this situation as playing a game. The actions in which teachers connect their knowledge about contexts, students, subject matter, didactics, and media are thus understood as ludic actions. Completely theoretically guided actions would require a full theoretical understanding of the situation, unlimited time to analyze the situation, the possibility to reject the action in case of any doubts and a complete knowledge of all participating persons. Obviously, this cannot be the case in education. Thus, educational actions perceived as artistic actions always carry aspects of Paidea [13].

With playful actions, teachers overcome the uncertainty gap – but they have to reckon they might lose the game. If they lose the game, the difference to serious actions shows up clearly: if teachers lose a round, they are not fired, they do not get bankrupt and, of course, they do not die – they just play another round of teaching and learning. And if they are good teachers, they try to play better next time.

At the risk of being boring, we have to repeat an earlier argument here: Dead people do not play. And, as you might have guessed, dead matter is not able to play. Thus, we are facing a problem similar to the initial one here: Computers do not play. Computers can be understood as toys [98], but machines are by no means able to play. Thus, computers cannot act as teachers, but they can be used to create playgrounds where teachers and learners play the game called teaching and learning. Since education in practice always has to take care of individuals, acting as a teacher which is an art form for Herbart. Thus, teachers are artists. And according to Schiller, artists do play.

From this point of view it is obvious, that teaching cannot be controlled or steered by knowledge that can be expressed in algorithms or data. One consequence is that designing an adaptive assistant system is not like designing an industrial robot for serious work. It's more like the creative design of an actual game, like the creation of a room where teachers and learners can play. This might be connected to the difference between game and play that is discussed in video game studies: "Play is an open ended territory in which make believe and world building are crucial factors. Games are confined areas that challenge the interpretation and optimizing of rules and tactics" [102]. Good games foster play, not work to earn one's living.

Games need to consider the rules of the game, while play is a free activity, where freedom is created by open up a make-believe world. Whether play in this sense actually happens cannot be predicted, but we can assume that toys are more likely to be played with than other objects [98]. The media didactical design of a game to be played by teachers and learners needs to consider basic educational problems and the possibilities of algorithms.

One example is the algorithms that have been developed by Brusilovsky et al. [10, 47]. The system developed by Brusilovsky et al. is used to teach Java. The algorithms developed by Brusilovsky and Hsiao allow for setting test question parameters. Questions are calculated. According to test results, links for students are adapted by showing colorful targets. This matches the concept of branched programming.

While this concept is a good idea for an introduction to a programming language, it is hardly possible to calculate variations of test questions that can be analyzed by an algorithm in other fields. Educational theories, for example, cannot be taught that way. Additionally, epistemological questions have not been considered by Brusilovsky et al., since differences among functional, procedural, and object-oriented programming are not taken into account. Different teaching methods are not considered at all. As a consequence, dynamic learning pathways cannot be created. The system offers all information for free navigation and considers the freedom of the learner this way.

But it cannot be transferred into other fields. And it is not possible to design learning pathways that do not contain tests that can be analyzed by an algorithm with this concept.

A second group of concepts applies algorithms that are based on the idea of artificial intelligence and suggest Intelligent Tutoring Systems. It is necessary to say a word on the term artificial intelligence from an educational point of view here. First, as we already stated for learning, intelligence in the term artificial intelligence has another meaning than intelligence in the term human intelligence. Second, human intelligence has a different meaning than the term thinking in philosophy, while thinking does not mean the same as understanding or learning in education. What is comparably clear is the definition of the term algorithm [58]. Considering the definition of algorithms it is clear, that neither understanding nor learning has anything to do with artificial intelligence.

Intelligent tutoring systems are based on algorithms. They are connected to the shift from batch processing to dialogue systems and problem solving theories. Additionally, extended computational power is used to Intelligent Tutoring Systems. The idea was first based on the concept for the General Problem Solver (GPS) [71], where the knowledge of problems and strategies to solve problems were separated. When the GPS failed for any relevant problem, the concept was replaced by expert systems [26]. The core architecture of the DENDRAL expert system [11] (knowledge base, explanation system, inference engine) became the starting point for SCHOLAR [12], which was built as a semantic network and based on the architecture of expert systems.

2.2 Computer Technology as a Medium in Teaching and Learning

Christian Swertz, Alexander Schmoelz, Alessandro Barberi and Alexandra Forstner

We understand media as things that are used as signs by human beings. With this broad term of media it is clear, that media need to be applied in all educational processes. This starts from the body in the medium of a gesture and reaches through oral communication to technical appliances like books, TVs or a computer.

From the different aspects of our media theory we would like to highlight one aspect here: Technical media are artifacts, and human beings express themselves in these artifacts. This describes a layer of communication, where the material of a medium is shaped in order to exchange ideas. This layer of communication has been highlighted by the Toronto School [48, 64]. Since using material in a medium is necessary for communication, this layer affects educational processes. In educational processes, it is impossible to avoid the bias of communication caused by the material layer, but necessary to choose or, if it comes to technical media, shape the material layer of the medium used.

Here, it is not possible to discuss criteria (like the interest in acceleration, individualization, etc.) for choosing or rejecting computer technology as a medium in educational processes. We just assume as obvious that it is possible to teach and learn with computer technology and describe the material layer of computer technology in order to inform the design of our tools.

Computers today are nearly always built as electrical universal Turing machines with a von Neumann architecture. This design of the material layer of the medium leads to a set of properties. One important property is that computers need to be programmed. Programming a computer is quite different from educational processes among human beings. The program determines the output of the computer, even if stochastic measures are used. That's why transferring information between computers has a very different meaning than learning in the field of education, as we already stated. We would like to add three observations to take a closer look at computer technology here.

The first observation is that the memory of human beings and the data storage of computer technology are quite different. Human beings can't forget. Of course, human beings do forget. But this is something that happens to human beings. It is not a competence. There is no "mastery of forgetting". To the contrary: The harder humans try to forget something, the better they remember it. Deleting data with computer technology is quite different: It can be executed on purpose. And it can be done sustainably.

The second observation is that there is an exact alignment among assembler commands and machine code in digital electric Turing machines. Since the meaning of machine codes in actually is the physical reality of the actual machine, there is no difference between symbols and reality for computer technology [58]. As René Magritte has illustrated with the words "Ceci n'est pas une pipe" on his famous picture "La trahison des images" [The Treachery of Images], this is not the case for human beings. For people, the relation of symbols and reality is problematic – to say the least. That's one reason why human beings become problems for themselves. Fortunately, this is not the case with computer technology.

These observations illustrate that the term "learning" signifies different concepts in computer technology and in education. The difference between these homonyms is the challenge when it comes to modeling didactic expertise with computer technology. With this challenge it is clear, that trying to replace teachers by computer technology is not an option. Machine learning and human learning cannot be converted; there is no jumper to close the open gap. That's why we consider computer technology as a valuable tool that can be used to design an adaptive assistant system for teaching and learning. From the media didactic point of view, the challenge is to create applicable algorithms and thus design the material substance that is used in the medium. Until now, we tried to elaborate some limits of the application of computer technology in education. With this in mind, we are going to discuss the history of Adaptive Assistant System in the next section.

2.3 The History of Adaptive Assistant Systems for Teaching and Learning

Christian Swertz, Alexander Schmoelz, Alessandro Barberi and Alexandra Forstner

When designing an adaptive assistant system for teaching and learning, a look at the history of these systems is informative. One of the interesting aspects is the impact of programing techniques that were fashionable at a time on the conceptualization of adaptive assistant systems.

If feedback is considered as a criterion for automated support in learning, the device presented by Pressey in 1923 was the first teaching machine In his paper, Pressey [77] stated that the device should not replace the teacher, but "make her free for those inspirational and thought-stimulating activities which are, presumably, the real function of the teacher". Skinner [92], who picked up Pressey's design as well as the foundation in the theory of Thorndike, also considered this limitation of machine support in learning. While Skinner applied feedback mainly as reinforcement in linear learning programs, Crowder's setting of intrinsic or branched programming offered a different feedback. His machine generated an individualized learning pathway [20] when a learner failed a test in a way that reflects the development of block-structured programming languages. The different learning pathways included additional content and explanations concerning the error, while individualization did not mean that the learner could make choices of his own.

This concept has become famous under the label programmed instructions [30] and is still used often. The concept is mainly based on tests that can be analyzed automatically. Today, this concept is called adaptive since current applications adjust the amount of tests, the available time for learning, the difficulty of questions, waiting times and hints while learning [53].

This first individual learning path component was extended by adaptive systems in the 1960s and 1970s [72]. Adaptive systems added a more sophisticated dialogue component to the programmed instruction systems and thus reflected the development of dialogue systems. This concept of adaptive systems is still developed today [34]. From a present-day perspective on programmed instruction, the connection between the actual machines and the theoretical concept is obscure on the one hand and many charges against behavioristic concepts are hardly sustainable on the other hand [52]. Maybe the second argument explains why behavioristic concepts are successfully applied in therapy today, but hardly in teaching and learning.

One example is the algorithms that have been developed by Brusilovsky et al. [10, 47]. The system developed by Brusilovsky et al. is used to teach Java. The algorithms developed by Brusilovsky and Hsiao allow for setting test question parameters. Questions are calculated. According to test results, links for students are adapted by showing colorful targets. This matches the concept of branched programming.

While this concept is a good idea for an introduction to a programming language, it is hardly possible to calculate variations of test questions that can be analyzed by algorithms in other fields. Educational theories, for example, cannot be taught that way. Additionally, epistemological questions have not been considered by Brusilovsky et al., since differences among functional, procedural and object-oriented programming are not taken into account. Different teaching methods are not considered at all. As a consequence, dynamic learning pathways cannot be created. The system offers all information for free navigation and considers the freedom of the learner this way. But it cannot be transferred into other fields. And it is not possible to design learning pathways that do not contain tests that can be analyzed by an algorithm with this concept.

A second group of concepts applies algorithms that are based on the idea of artificial intelligence and suggest Intelligent Tutoring Systems. It is necessary to say a word on the term artificial intelligence from an educational point of view here. First, as we already stated for learning, intelligence in the term artificial intelligence has another meaning than intelligence in the term human intelligence. Second, human intelligence has a different meaning than the term thinking in philosophy, while thinking does not mean the same as understanding or learning in education. What is comparably clear is the definition of the term algorithm [58]. Considering the definition of algorithms it is clear, that neither understanding nor learning has anything to do with artificial intelligence.

Intelligent Tutoring Systems are based on algorithms. They are connected to the shift from batch processing to dialogue systems and problem solving theories. Additionally, extended computational power is used to Intelligent Tutoring Systems. The idea was first based on the concept for the General Problem Solver (GPS) [71], where the knowledge of problems and strategies to solve problems were separated. When the GPS failed for any relevant problem, the concept was replaced by expert systems [26]. The core architecture of the DENDRAL expert system [11] (knowledge base, explanation system, inference engine) became the starting point for SCHOLAR [12], which was built as a semantic network and based on the architecture of expert systems.

In this concept, limitations were hardly considered, and learners could only barely make their own choices. Despite the effort invested in ITS there are hardly actually working systems available or real world applications reported. ITS seems to have failed due to the high effort necessary to develop such systems and the lack of theoretical foundations [91]. From our perspective, considerably basic educational problems like the theory-practice-transformation were not considered in the design of ITS.

In the last years, the successful application of recommender systems in marketing led to the idea of transferring the concept of those systems in the didactic field [23]. This often takes place in the context of informal learning processes [62]. The concepts seem to be related to constructivistic learning theories, while explicit references are rare. While most of the suggested systems are in the early stages of development, the expectations are high. At least, these expectations appear to be similar to the systems discussed before. Since the difference of marketing and didactics is not considered yet for recommender systems, similar problems can be expected as well.

With systems for programed instruction, intelligent tutoring systems, adaptive learning environments, and pedagogical recommender systems concepts for automatic educational reasoning have been developed. These systems haven been developed for many decades. Despite the effort invested there are hardly actually working systems available or real world applications reported. Intelligent tutoring systems seem to have failed due to the high effort necessary to develop such systems and the lack of theoretical foundations [90]. This might be connected to one concept all the systems developed so far share: Developers assumed that learning is a formally describable and controllable process. Fortunately, this assumption is wrong.

Neither the General Problem Solver nor the Intelligent Tutoring Systems that were based on the General Problem Solver were useable or successful. [90]. This applies to current systems that are based on the same concept too. One example is the concept developed by Bredweg and Struss [9]. Based on an overview on qualitative reasoning they show that the strength of qualitative reasoning is the consideration of causality. They argue that this consideration of causality is a strength of the approach, since causality is essential for model building in scientific thinking. As a conclusion, they focus on the presentation of cause-and-effect-chains in artificial intelligence algorithms. This presentation is turned into educational objectives. Learners should learn the cause-and-effect-chain thinking by modeling causal relations with cybernetic qualitative intelligent algorithms.

That way, only one epistemological concept is considered. Unfortunately, this is not explicit – the epistemological position is not discussed by the authors. A reference to the theory of modeling [94] order representation theory [104] is missing as well. By doing so, the freedom of the learner that is connected to choosing an epistemological position is neglected. Since the necessity to reflect scientific methods is neglected as well, the approach can hardly be understood as scientific thinking. It is focused around the idea of an operative cybernetic control system. Since such a system is based on algorithms, it creates a self-contained world [58] and thus the illusion of a predictable and known future.

Another approach is algorithms that conduct tests of learning styles and present content accordingly. One example for a study like that has been published by Lehmann [60]. It is based on the learning style inventory developed by Kolb [56]. Content has been prepared for a learning cycle that allows for the consideration of learning styles [60]. Learners have been tested. They were randomly spread on treatment groups so that the content was presented in a way optimized due to the results of the learning style test.

This study shows several problems: First, the research by Lehmann was based on a small incidental sample from a small basic population. The results can thus not be generalized. Second, there were hardly any relevant results. This is not astonishing, since designing content based on learning style inventories, that is on a perspective based on averages was not successful before [49, 50]. From a didactic perspective this was expected, since learning style theories do not take into account that learners do not learn content only, but also learn to learn [99], as we already stated.

The first adaptive systems have been developed in the 1960s and 1970s [73]. One contemporary example for an adaptive system is the approach suggested by Martens [63]. Marten's Tutoring Process Model (TPM) is a formal approach to the design of Adaptive Tutoring Systems. A prototype based on the concept has been developed. The prototype is not available anymore and has not been used in other projects. This is a faith shared by many prototypes in the field of didactics [90].

Martens defines the tutor model as $TPM =< C, \; LM, \; show, \; enable >$ with $C =< Q, \; A, \; q0, \; F, \; B, \; \delta, \; select, \; allow >$ and

Q: finite set of states
A: finite set of actions
$q0 \in Q$: start state
$F \subset Q$: finite set of final states
B: finite set of bricks
δ: state transition function
select: select brick function
allow: select action function.

With this definition, building adaptive menu systems becomes possible. A learner model can be considered formally. Only elements to inform and to interact are considered as building blocks. Cooperations are missing. This limits the possibilities of the model. Similar limits exist in other models [14].

It can be concluded that educational problems are not sufficiently considered in the discussed approaches. The algorithms are limited to isolated cases and small content areas like Mathematics, Programing and Languages. In nearly all cases only standardized parts which are located at the beginning of curricula were considered. Many algorithms that are developed today fall behind the approaches discussed. They only use simple versions of programed instruction. In some cases successful applications in certain subject didactics have been created. But none of the approaches designs the leeway in the communication among teachers and learners by considering media didactics.

Another point is that computer technology is neither capable of creating art nor able to play. Thus, computer technology can never replace teachers. Maybe it can simulate learners that make teachers happy but this has hardly been researched yet.

2.4 Conclusions

The argumentation in the first sections leads to a different status of Adaptive Assistant Systems. While previous concepts tried to replace teachers, we try to create tools for teachers. These tools are intended as toys that suggest teachers to play with their teaching methods and the media they apply. If teachers play with teaching methods and media and offer differences and varieties, they again open up a playground where students can learn while playing with these teaching methods and media.

Based on this perspective, designing an Adaptive Assistant System places us in the position of designing tools for creating games. These tools can be used to create a playground for teachers that act as artists who create games for learners. Pictorially we create brushes and colors that are used by teachers to paint pictures that are shown to the learner. Thus, the challenge is to design tools for the creation of teaching and learning processes that open up spaces for creative actions. The fact that the contradiction between compulsory rules and open creativity is solved without any problem, while actually playing games and shows in turn that the association of gaming for teaching and learning is suitable.

It is obvious that a supplier of brushes and colors has hardly any control about the created artwork that will be presented to the audience. The only thing he can assume is that the color will be present in the artwork in which form ever. This is considerably the case if you think about something like audience participation in non-scripted performance art. Since we consider Adaptive Assistant Systems as tools for teachers and not as a replacement for teachers and according to Herbart acting as a pedagogue is an art form it does not make any sense for developers of Adaptive Assistant Systems to even try to control learning environments and learning outcomes above all. A consequence of this is that learning outcomes cannot be applied as a measurement for a successful design of an Adaptive Assistant System. Still, this measure has been applied as the only measure in recent decades. Thus, it is necessary to develop new criteria for the success of Adaptive Assistant System. We assume that human beings do have an own free will, need to live in a community, and need to be understood as decision making agents. Freedom and the open future are considered as essential. Starting with this assumption, the possibilities and limitations of computer technology in teaching and learning have to be considered.

If the possibilities and limitations are considered, computer technology can be used as an assistant system for teachers and learners. Since computer technology needs to be programmed, programmers have to be considered as teachers that set up the setting in which other people teach and learn. In this respect, their actions can be understood as a kind of policy making for teaching and learning. Designing, implementing, and deploying software for teaching and learning is an educational act. Since the software is usually used as it is, software is an instrument to claim power.

In this respect, the balance of force and freedom as a basic educational problem needs to be considered. The developing freedom of learners has to be taken into account. From an educational point of view, the software for teaching and learning has to be designed in a way that suggests and allows learners to develop their freedom. This can be done by offering learners' tools to increase control on their learning processes. Of course, this is a claim to power again and refers to the basis dialectic of freedom and force that is inevitable in education.

Instruments that support learners' control can consider the content and the learning process. Since our project aims at a content independent software, the learning process can be taken into account only. To do so, data about the learning process have to be collected and analyzed. The results have to be turned into recommendations for the learner. If the recommendations reproduce teachers input only, they are pointless. Adaptive Assistant Systems become relevant for education if they support creative behavior by the learner and thus support learners to create their own way of learning.

In INTUITEL, this is applied to Learning Pathways and Feedback. Learners should be supported in choosing from different learning pathways and in creating their own learning pathways. Feedback can be created by considering learners earlier behavior and by considering other learners' behavior. This again can be used to create recommendations only. It has to be possible that learners deviate from recommendations issued by the software.

Finally, the freedom of teachers has to be considered as well. It has to be possible to express different content structures and arrange content according to different learning theories. At the end, it is necessary to include the possibility for teachers to try to force learners to learn in a certain way, while we cannot predict which way this will be. Thus, a structure to allow teachers to express different ideas of teaching is necessary too. These requirements can be matched by reasons that are applied to dynamic hypertexts which are based on a didactic ontology and the collection of data about teachers and learners. In other terms: INTUITEL is about ontologies and reasoning in education.

3

The INTUITEL Approach:
Foundations and Design

This chapter explains the modeling in the INTUITEL approach. It picks up the idea of using ontologies and reasoning to model didactic expertise from the previous chapter. The concept of the ontology of pedagogies, the idea of learning pathways and the learner model are described. Didactic factors are introduced. The model how they are used to deduce recommendations and feedback in real-time is developed. In the fourth part we describe the software architecture and in the last part the data model and communication standard is explained. The decision for semantic technologies and the OWL2 specification is justified.

3.1 Pedagogical Ontology and Reasoning

Christian Swertz, Alexander Schmoelz, Alessandro Barberi
and Alexandra Forstner

An ontology needs to be consistent from a technical perspective [35]. In contrast, teaching and learning is inconsistent due to the artistic nature of educational actions. Thus, the challenge is to build an inconsistent consistency, which is an ontology that opens up a consistent room which is necessary to meet the logical structure of computer technology and that allows for the creative design of teaching and learning processes. The gap that is indicated by this contradiction can be filled by teachers and students when playing with the system.

We suggest providing a meta-data system, a learner model and a reasoning engine as tools to create learning environments. The meta-data system allows teachers to describe different possibilities to learn certain content. It can be formulated logically in an ontology in the Web Ontology Language.

The flexible elements are circled around learning pathways. The learning pathways, defined as relations between concept containers (CCs), between knowledge types, and between media types can be altered by teachers and by learners. If a teacher, for example, prefers other steps than suggested by a didactic model, he can mix those steps with steps from other pathways or create steps. While doing so, he plays with the teaching and learning models that were applied while creating the meta-data system. Some basic teaching and learning models are suggested (inquiry-based learning and multistage learning), but the teacher neither has to follow these models nor to apply these models at all. He is always free to create his own learning pathways and offer them to the learner.

Thus, the meta-data system allows teachers to play with various teaching models. Still, he has to describe his learning material with this meta-data. In his game he still uses the meta-data system, but as a toy. Since the teacher uses the meta-data system an automatic reasoning engine is still able to react on the results from teachers play. Since the learning material and the meta-data developed by the teacher are offered to learners they can use these to play too. If for example a teacher creates a learning sequence, the learner can learn the material backwards or in any creatively created order. This order can automatically be identified, converted in a personal learning strategy and applied to further material. Since the different learning pathways and the descriptions are offered to the learner, a flexible room is created where learners can play with learning models.

Understanding teaching and learning (at least partly) as play and computer technology as a toy used to create a playground sheds some light on the position that is taken when creating a pedagogical ontology for machine support in didactic practice: we are creating a game for people who play a "create a game" game. With computer technology, the playground can be best modeled by an ontology [69]. This form of a semantic network specifies the rules of the game. In order to do so, it is necessary to open up different possibilities for expressing ideas of teaching and learning creatively. Still, some rules have to be set when creating games. In order to keep the possibilities open, these rules can be developed from an analysis of computer technology as a medium, since the properties of a medium applied in teaching and learning always limit the possible actions.

The consistent part of the ontology we propose consists of a three level meta-data system for learning objects [66]. Learning Objects include instructional scaffolding such as learning objectives and outcomes, assessments, and other instructional components, as well as information objects [67]. We accommodate the levels of learning objects by using three types of Learning Objects: (i) Knowledge Domain (Course Level), (ii) Concept Container (Lesson Level), and (iii) Knowledge Objects (KOs; Content Level). The term Knowledge Domain refers to a certain amount of knowledge, which is defined by a specific curriculum, syllabus and/or course requirements.

One CC contains one instructionally framed concept within a Knowledge Domain. A CC is a container for one or more KOs. A KO is an item of knowledge, which typically corresponds to about one screen page of content and to an estimated learning time of 3–10 min for the average learner. A KO might contain learning content as well as learning activities such as a discussion in a forum, an assignment where a video has to be handed in or reading an explanation. KOs are described by a pedagogical knowledge type and a media type. CCs and KOs can be connected by relations.

In order to support different learning pathways, a vocabulary has been developed. The vocabulary for the CCs is intended to express the structure of the knowledge domain. It considers the hierarchical relations has child, has parent, and has sibling as well as the chronological relations is before, is after and is beside. The vocabulary for the knowledge types is intended to express pedagogical concepts. The vocabulary for the media types is also intended to express pedagogical concepts.

3.1.1 Learning Objects

The INTUITEL ontology is based on the concept of learning objects. Learning Objects include instructional scaffolding such as learning objectives and outcomes, assessments, and other instructional components, as well as information objects [67]. INTUITEL will accommodate Metros dimensions of learning objects by using three types of learning objects:

1. Knowledge Domain (Course Level)
2. Concept Container (Lesson Level)
3. Knowledge Objects (Content Level)

Thus, learning objects contain learning objects of different object types (see Figure 3.1).

The term knowledge domain in general refers to the part of the world investigated by a specific discipline. In INTUITEL, the term knowledge domain refers to a certain amount of knowledge, which is defined by a specific curriculum, syllabus and/or course requirements. In INTUITEL four partners

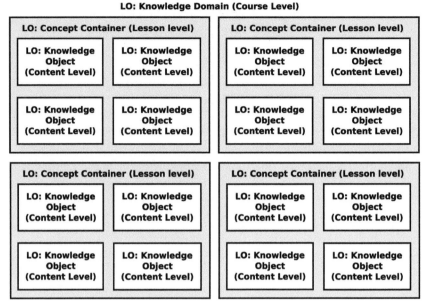

Figure 3.1   Learning object hierarchy in the pedagogical ontology of INTUITEL.

(IOSB[1], URE[2], UVA[3], and UVIE[4]) will provide four cognitive models of four different knowledge domains, which correspond to the different example courses of INTUITEL. Knowledge Domains consist of several concept containers. Course is a synonym for knowledge domain. Knowledge Domains have a title and consist of knowledge containers.

One CC contains one instructional scaffolding concept within a knowledge domain. CCs are part of a knowledge domain. CCs are linked by typed relations within the knowledge domain. CCs are assembled and structured corresponding to the logic of different pedagogical CC models that are derived from learning pathways and expressed by the typed relations. Concept containers have a title, typed relations to other CCs, and are part of a knowledge domain.

Knowledge objects contain about one screen page of content and correspond to a learning time of 3–10 min. A KO covers mainly one knowledge type and one media type. The content of a KO can be anything like:

---

[1]Fraunhofer Institute of Optronics, System Technologies and Image Exploitation.
[2]Universiy of Reading.
[3]University of Valladolid.
[4]University of Vienna.

- a discussion in a forum (knowledge type: discussion and media type: text),
- an assignment where a video has to be handed in (knowledge type: hand in assignment and media type: video),
- reading an explanation (knowledge type: explanation and media type: text).

Knowledge objects are assembled and structured corresponding to the logic of different pedagogical knowledge type models and media type models that are derived from learning pathways. KOs have a learning time, a knowledge type, and a media type, are part of a CC and consist of content.

3.1.2 Vocabulary of Knowledge Types

Knowledge Types are due to didactical requirements. However, this structure of knowledge must be always seen as preliminary, because it can only be structured according to the goals of the knowledge type structure. Knowledge types are structured by means of the function within the learning process. This is the didactical goal of the organization of knowledge for the learning process. Functions within the learning process are presentation (receptive knowledge), trial (interactive knowledge) and communication (cooperative knowledge).

3.1.2.1 Receptive knowledge types

Receptive Knowledge Types (e.g., Orientation, Explanation) contain media for presentation. Within the media, the knowledge is displayed but without changing the presentation because of the media. The presentation is static. The learner is receiving the knowledge but is not active beyond that. Receptive knowledge may be orientation, explanation or source knowledge.

Orientation Knowledge gives orientation in one field. Knowledge is orientation knowledge, if it is naming and relating the field with other knowledge and if it can be connected to previous knowledge of the learner. This knowledge is represented in terms of: facts, history, news, log, overview, knowledge map, abstract, and scenario.

Knowledge is an explanation, if it gives reasons for representations or claims. An explanatory statement is necessary, because representations can always be different. An explanatory statement for a representation names the method, which is used by the representation. Explanatory Statements are arguments, examples, descriptions, interviews, comments, definitions, exemplifications or ideas/tips.

Sources answer the question as to find information. If a person is in possession of sources, he/she can answer the question "where to find knowledge".

Therefore, the sources must be published and known. References on sources are made through indications of sources. Sources differ in types. The important types of sources are link lists, lists of literature, and download (which can be addresses or archives).

3.1.2.2 Interactive knowledge types

Learning items with interactive knowledge contain knowledge, whose presentation is influenced by the activity of the learner. The activity of the learner within the learning process is very useful if knowledge can be learned in an explorative way and if this knowledge can be proved in action or the knowledge is tested within an assignment.

3.1.2.3 Cooperative knowledge types

A didactical cooperation is a communication between humans, in which they work together on a certain topic in order to understand each other above expertise. Cooperative knowledge items are essential in order to react on unscheduled required knowledge. Cooperative Knowledge can be procured planned or spontaneous.

3.1.3 Media Type Vocabulary

3.1.3.1 Communication

Communication Media Types are described as tools for people to communicate directly with each other. In this list are only media types which are used online within networked computer technology. This may – for example – comprise chats, audio-conferences, video-conferences, and shared applications.

3.1.3.2 Interaction

An example for interactive media types are forms, where structured documents with blank spaces have to be filled out for further processing through a LMS. These blank spaces, which have to be filled out by the learner can be check boxes, radio buttons, lists, etc. Another example is interactive videos, where the user can at least interact through stop-and-go-functions with the computer. It would get better if the learner could also influence the plot of the interactive video.

3.1.4 Learning Pathways

Theoretically, learning pathways can be deduced from the logical structure of a knowledge domain that is expressed in the typed relations. In practice this would require a very well-written hypertext with precise typed and set relations. Unfortunately, authors tend to make little mistakes considerably in larger courses. Additionally, authors would need to know a lot about the logic of the adaptive assistant system in order to predict the outcomes that will be created based on their input. Finally, the automatic deduction of learning pathways would restrict authors to the pathways that are predefined in the system. Since there are hundreds of models for teaching and learning available and new ones created very often, this restriction does not make much sense. It would just create a tendency to undermine the theory-practice transformation competence of teachers. And that should be avoided.

That's why in INTUITEL the simple possibility to set different learning pathways among the same learning objects is considered. The learning pathways have to be set as directed acyclic graphs. No further restrictions apply. In addition to setting the learning pathways the teachers have to create a description that supports the learner in the pathway selection. Since Concept Containers and KOs are distinguished, Macro-Learning Pathways among CCs and Micro-Learning Pathways among KOs are possible. The Macro-Learning Pathways are on the level of the Content Container within one Knowledge Domain. The Macro-Learning Pathways describe how the learner might proceed within one Knowledge Domain. Within one Knowledge Domain, there can be more than one CC. These CCs are assembled and structured by learning pathways. The pathways are expressed by typed relations. In an example Knowledge Domain four Macro Learning Pathways have been used by the teachers:

- Chronologically from old to new
- Chronologically from new to old
- Hierarchically top down
- Hierarchically bottom up

Concept containers have a title; typed relations to other concept containers, and are part of a knowledge domain. The Micro-Learning Pathways are on the level of the KOs in one CC. The Micro Learning Pathways describe how the learner might proceed within one CC.

Within one CC, there can be more than one KO. If there are many KOs, they are assembled and structured by learning pathways. In INTUITEL there were three Micro Learning Pathways created by teachers for testing purposes:

- the MultiStage Approach
- the Inquiry-Based Learning Approach
- the Programed Instruction Approach.

An example metadata set for one KO is listed in Table 3.1.

Table 3.1   Example meta-data for a KO about Comenius

Meta Tag	Value
ID	KO_ComeniusOrientierungVideo LMS
Project	INTUITEL
Licence	Creative Commons Attribution Non-Commercial Share-Alike 3.0 Unported License
Author	Christian Swertz
Date	4.9.2013
Knowledge Domain	General Didactics
ContentContainer	Comenius
KnowledgeType	OrientationReceptive
Media Type	VideoReceptive
MicroLearningPathways	KO ComeniusOrientierungVideo isMoreConcreteThan KO ComeniusOrientierungText
Level	All
European Qualification Framework	LearnerEqfLevel6
EstimatedLearningTime	00:07:00
Suitable For Blind	Learner Is Not Blind
Suitable For Deaf	Learner Is Not Deaf
UnableToSpeak	All
Age	LearnerIsChild
Gender	All
Lang	De-de
Screen Minimum	320X240
Screen Recommended	640X480
Subtitle	None

This meta-data describe a seven minute video about Comenius. For an improved readability, only one Micro-Learning Pathway is reproduced here. These metadata are used as an input for the learning analytics integrated in INTUITEL. The results of learning analytics are used for adaptations, recommendations, and feedback.

3.2 Learning Analytics by Didactic Factors

Christian Swertz, Alexander Schmoelz, Alessandro Barberi,
Alexandra Forstner, Alexander Streicher and Florian Heberle

As inputs for the learning analytics component of INTUITEL, three sources are available:

1. Observations of teachers and learners
2. Input of teachers
3. Input of learners

Observation data exist in the form of log files where it is recorded which learning objects were when accessed by teachers and learners. The main input from teachers is the meta-data described previously. Input from learners is either derived from profile data that is available from learning manage ment systems or from answers learners gave to requests for input from the INTUITEL system which is called TUG messages.

Since it is pretty difficult for learners to analyze raw data while learning takes place, it seems appropriate to offer some results from learning analytics to the learner. Unfortunately, we do not know beforehand which results will be relevant to the learner, but have to prepare analytics before the learning takes place. Thus, the results should only be turned into recommendations to the learner.

If for instance observation data show that the last login was a fortnight ago, it might make sense to recommend a repetition of the last topic instead of continuing with the next one. Another example is the recommendation for a learning pathway based on the age and gender of the current user:

"This course can be learned by multi-stage learning, inquiry based learning or programmed instruction." Other learners of your age and gender preferred programmed instruction. Which model do you prefer? Unfortunately, it is not known yet which Feedbacks are useful. Since this is an empirical question, the system needs to be designed in a way that allows for subsequent adaptations. That's why the rules to create feedback will be written in OWL and not as software.

A Didactic Factor is a compound of a number of data items from INTUITEL in a way so that the combination of them describes a fact that is relevant for the recommendation creation. They are the fundamental building blocks of the Rating Factors, which are used to evaluate the suitability of KOs. For this purpose, everything that is available in the whole collection of INTUITEL data, meaning the SLOM meta-data, the Learning Pathways and especially the learner-specific information (e.g., the learning history as contained in the INTUITEL logs) that are stored or collected just-in-time from the LMSs, can be used.

From a technical perspective, a Didactic Factor is an OWL class which contains its own textual description. It furthermore also links to a Java class, containing its Transformation Rule. These are the instructions that specify in which combination of input data the respective Didactic Factor is valid. This combination of OWL and Java allows a very high flexibility regarding their specification, because all features of a high-level-programming language can be used. This especially also includes functionalities that would not be available in an OWL-only solution, as, for instance, mathematical methods to calculate the ratio between two values.

As seen from a reasoning point of view, the basic task of a Didactic Factor is to combine information in a way that allows its usage in context of an OWL-reasoner. These complex software modules have foundationally different intentions than programming frameworks like, for instance, Java or .Net. Instead of iteratively executing program code to produce various results, OWL-reasoners are specialized on testing the consistency of statements and the identification of relations between entities. By drawing conclusions on a data set (i.e. an ontology), a reasoner can deduce statements that, for instance, allow to determine whether a CC is fitting for a certain learner's Learning Pathway (LP). The Didactic Factors are especially relevant in this process because natural or real numbers are problematic in that context. This entails that INTUITEL needs to reformat the input in a way that is compatible with such a system. One aspect of the Didactic Factors is consequently to transform the non-nominal values into a nominal form (e.g., by transforming the continuous value 5 into the categorized statement "medium"). There are four fundamental forms of Didactic Factors:

1. Trivial statement: The most basic realization of a Didactic Factor is the n: 1 relaying of input data. This means that certain data items are combined and translated into a format that is compatible with the Engine. (example: gender as male or female)
2. Trivial input combination with grading: Different nominal data items can be connected to create a combined statement that entails some kind of grading. (example: connection type as slow, medium, and fast)
3. Complex statement: A more complex use case for a Didactic Factor is the discretization of numerical values into a nominal one (example: noise level in DB is expressed as quiet, tolerable, and loud)
4. Complex input combination with grading: The combination of different (kinds of) input values through, e.g., mathematical functions, can also result in graded Didactic Factors.

In the table below, we provide the list of the Didactic Factors that have been developed in INTUITEL. This list does not claim to be complete or that the respective items are final, since there is no evidence for useful factors available yet. This list will nevertheless give a detailed overview about aspects that might be relevant for the selection of suitable Learning Objects.

#	Didactic Factor	Description
01	Knowledge actuality	Ranking of time between now and the last learning session.
02	Course-focused KO learning speed	Ranking of learning time the learner on average differs from the estimated learning time in contrast to the same measure of the other course participants
03	Learner-focused KO learning speed	Ranking of learning time the learner on average differs from the estimated learning time of completed KOs of this session in contrast to same measure over all KOs over all sessions.
04	Course-focused filtered KO learning speed	Ranking of learning time the learner on average differs from the estimated learning time in contrast to the same measure of the other course participants when only having a look at KOs that have the same KT and MT.
05	Learner-focused filtered KO learning speed	Ranking of learning time the learner on average differs from the estimated learning time of completed KOs of this session in contrast to same measure over all KOs over all sessions when only having a look at KOs that have the same KT and MT.
06	Course-focused session length	Statement about the average session length as compared to the average session length of other course participants.
07	Learner-focused session length	Statement about the current session length as compared to the average session length of the learner.
08	Time exposure	Comparison between the amount of time the learner and the other course participants spent on the course.
09	Learning Pathway permanence	Ranking of the amount of KOs the learner completed on the current LP combination in contrast to the same measure for the other course participants.
10	Recent learning pace	Comparison of the actual learning time the learner needed for the last 10 KOs in contrast to the estimated learning time.
11	Session learning pace	Comparison of the actual learning time the learner needed for the KOs in this session in contrast to their estimated learning time.
12	Course-focused LP usage type	Statement about the LP usage as measured on the learners pathway switches and the switches of the other course participants.
13	Learner-focused learner type	Statement about the LP usage as measured on the learners pathway switches.
14	Course-focused learning success	Success of the learner regarding scores in contrast to the other course participants.

(*Contnued*)

Table   Continued

#	Didactic Factor	Description
15	Learner-focused learning success	Success of the learner regarding scores in contrast of the own score history.
16	Course-focused KO repetition quantity	Comparison of the number of repeated KOs with the number of repetitions of the other course participants.
17	Learner-focused KO repetition quantity	Comparison of the number of repeated KOs in the recent KO history and the average of repeated KOs.
18	Course-focused CC repetition quantity	Comparison of the number of repeated CCs with the number of repetitions of the other course participants.
19	Learner-focused CC repetition quantity	Comparison of the number of repeated CCs in the recent KO history and the average of repeated CCs.
20	Course KO completion	Statement about the coverage of the course regarding the completion states of KOs.
21	Course CC completion	Statement about the coverage of the course regarding the completion states of CCs.
22	CC KO completion	Statement about the coverage of the current CC regarding the completion states of the connected KOs.
23	Course-focused KO completion tendency	Comparison of the learners and the other course participants ratio of completed KOs in contrast to the uncompleted ones of the session.
24	Learner-focused KO completion tendency	Comparison of the earners and the other course participants ratio of completed KOs in contrast to the uncompleted ones of the session.
25	Course-focused MT preference	Statement about the MT preference as measured on all course participant selections.
26	Learner-focused MT preference	Statement about the MT preference as measured by the learners learning history.
27	Course-focused MT dislike	Statement about the MT dislike as measured on all course participant selections.
28	Learner-focused MT dislike	Statement about the MT dislike as measured by the learners learning history.
29	Course-focused KT preference	Statement about the KT preference as measured on all course participant selections.
30	Learner-focused KT preference	Statement about the KT preference as measured by the learners learning history.
31	Course-focused KT dislike	Statement about the KT dislike as measured on all course participant selections.
32	Learner-focused KT dislike	Statement about the KT dislike as measured by the learners learning history.
33	LP leaving position	Statement at which point (in the sense of completed LOs) the learner leaves a LP.
34	Course-focused learning efficiency	Ranking of how much time the learner needs to complete a KO in contrast to the time the other course participants needed for it.

Table Continued

#	Didactic Factor	Description
35	Learning attention	Statement about how much attention the learner pays to the content as measured by an eye-tracking device connected to the LMS.
36	Blindness	Statement if the learner is blind.
37	Deafness	Statement if the learner is deaf.
38	Gender	Statement about the learners gender.
39	Age	Statement about the learners age.
40	EQF Level	Statement about the learners European Qualification Framework (EQF) level.
41	Learner Level	Statement about the course specific level of knowledge the learner possesses.
42	Device resolution	Ranking of the relative resolution of the device the learner uses to access the LMS.
43	Connectivity level	Ranking of the connectivity between the learners access device and the LMS.
44	Noise level	Ranking of the environmental noise level of the learner.
45	Learning Environment	Statement about the type of environment the learner is currently located at.
46	Learning Velocity	Ranking of the time the learner needs to successfully complete Learning Objects.

As stated above, these factors need to be transformed into statements that can be computed by a reasoner. To give an example, let us assume that the estimated learning time is 3 min and the actual learning time was 2 min, and 30 s. Let's further assume that the transformation rule differentiates five cases:

1. Estimated time actual time > 2 min $\Rightarrow$ No rating
2. Estimated time actual time < 2 min AND > 1 min $\Rightarrow$ fast learner
3. Estimated time actual time < 1 min AND > -1 min $\Rightarrow$ normal learner
4. Estimated time actual time < -1 min AND > -2 min $\Rightarrow$ slow learner
5. Estimated time actual time < -2 min $\Rightarrow$ No rating

In the present example, this would result in the statement that the learner is a normal learner. Please note that this is only an example. There can be arbitrarily many combinations of input values, but only a subset of them is pedagogically meaningful and exact enough. If, for example, the estimated learning time is quite high (e.g., hour which is non-compliant to the INTUITEL guidelines), completing the KO more than one minute earlier or later is certainly common. Thus, specifying well-engineered rules is an important factor regarding the accuracy of INTUITEL.

Concluding this section, the following examples explain the transformation rules for three of the above mentioned Didactic Factors. Please note that due to a better readability, standard deviation is denoted as *s*. For a full definition of all transformation rules of the 46 Didactic factors, refer to the according Deliverable 3.2 [25] of the INTUITEL project.

Transformation rule for DF "Course-focused KO learning speed"

Input:
lAvgLT = learners average learning time of recent KOs
oAvgLO = others average learning time

Output:
KoSpeedFast, KoSpeedSlow, KoSpeedNormal

Transformation rule:

```
if (lAvgLT > oAvgLT + s)
    output = KoSpeedSlow
else if (lAvgLT < oAvgLT - s)
    output = KoSpeedFast
else
    output = KoSpeedNormal
```

Transformation rule for DF "Course-focused filtered KO learning speed"

Input:
lCouples[] = Learners average difference between actual and estimated learning time of KOs, which is differentiated into KT and MT couples (only the three topmost types each),
oCouples[] = Others average difference between actual and estimated learning time of KOs, which is differentiated into KT and MT couples (only the three topmost types each).

Output:
FilteredKoSpeedFast, FilteredKoSpeedSlow, FilteredKoSpeedNormal

Transformation rule:

```
For each couple {
```

```
    if (couple not null for learner) {
        lAvg += learner s value for couple
            oAvg += others value for couple
          }
}
lAvg /= count of not null couples
oAvg /= count of not null couples
if (lAvg > 110\% of oAvg)
    output = FilteredKoSpeedSlow
else if (lAvg < 90\% of oAvg)
    output = FilteredKoSpeedFast
else
    output = FilteredKoSpeedNormal
```

Transformation rule for DF "Learner-focused learning success"

Input:
scoRec = Recent average learner score
scoGen = General average learner score

Output:
SuccessBetter, SuccessStable, SuccessWorse

Transformation rule:

```
if (scoRec > scoGen + s)
    output = SuccessBetter
else if (scoRec < scoGen - s)
    output = SuccessWorse
else
    output = SuccessStable
```

3.3 Learning Progress and Learning Pathways

Alexander Streicher and Florian Heberle

The central information mediating component of INTUITEL is the Learning Progress Model (LPM) module, which connects all other components. It acts as a preliminary stage for the INTUITEL Engine. By providing functions to perform transformations of learner scores, history, and pedagogical as well as domain knowledge into a position within a cognitive space, the LPM prepares the data to be usable by the semantic reasoner in the INTUITEL Engine.

In order to achieve this, it relies on multiple data sources which carry data about the users. First, there is the data about the learner as represented in the LMS. Secondly, the pedagogical and domain knowledge in form of the Pedagogical Ontology and the respective Cognitive Models, provide information about the actual learning content and how to guide a learner through the learning material. And in the third place, a user model of internal INTUITEL relevant user data like session data or the navigation history. The specific tasks of the LPM regarding the recommendation creation process can be summarized with three core themes:

1. Data aggregation
2. Data storage
3. Data transformation

To complete its range of functions, the LPM includes components to trigger direct user feedback before the reasoning process starts, determine the most suitable Learning Pathway (LP) and create the optimized data basis for the Reasoning Engine. Explicitly not task of the LPM is finding suitable Learning Objects (LO) for a learner. This is solely task of the INTUITEL Engine. The LPM is only indirectly involved in this process since it provides and edits the relevant information to be later used by the INTUITEL Engine.

3.3.1 INTUITEL Recommendation Process

There are multiple components that are part of the recommendation creation process and the individual tasks need to be distributed between them. The LPM is one of these components and the following descriptions show which tasks are conducted by the LPM and which tasks the other modules take over. This also depicts how the LPM is positioned in the INTUITEL overall system design.

The functional process has three stages, as depicted in Figure 3.2. There firstly is the stage of data preparation where the LPM collects and reformats the data for the INTUITEL Engine. Secondly, the reasoning process is conducted to determine the elements in the different result sets. Thirdly, the results are sorted and reformatted to be in a format that is compatible with the rest of the INTUITEL system. Also, learning recommendation messages are created to guide the LMS user.

3.3.2 Comparison to Real-Life Tutoring

Speaking in terms of a real-life learning scenario, INTUITEL is a workflow engine to process well established learning methods, in particular by deducing relevant and individual recommendations for learners. Without loss of generality, Table 3.2 outlines the analogy between the decision process of a real teacher and INTUITEL.

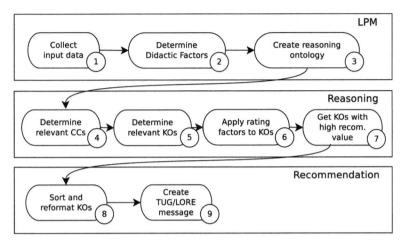

Figure 3.2 Simplified functional procedures for the recommendation creation in the back end.

Table 3.2 Comparison between learner and INTUITEL behavior in learner counseling

Pedagogical Aspect	INTUITEL Analogy
A human teacher senses when a learner needs assistance. The sensing process may rely on subconscious experience based processes as well as involve cognitive reasoning	Via the LMS-interfaces, INTUITEL registers and protocols the current and precious states of the learner. The LPM serves as the sensing instance, either by simple rules based inference or by more complicated ontology based reasoning.
In looking at test results and carrying out a spoken dialog, a teacher tries to assess the learner's current needs.	Usage of personalized information INTUITEL has collected for each learner and request of up-to-date information via USE or TUG dialog.
By applying personal didactic/domain knowledge or by intuitive reaction, the teacher forms an opinion about the learner's situations.	Application of transformation rules to create the different Didactic Factors and their integration in the learner-specific Learner-State-Ontology.
The teacher evaluates the situation on a topical basis and aligns it to his or her personal knowledge about the topic.	With the Cognitive Model stored in the SLOM file, INTUITEL draws conclusions about necessary alignment of the learning process.
The teacher ponders which advice fits best.	Reasoning process in the INTUITEL Engine to select suitable KOs and/or the decision which natural language feedback should be given.
Teacher gives the learner an advice which fits his/her current situation best.	Recommendation of specific content via the LORE interface and/or natural language messages via the TUG interface.

3.3.3 Reflex Reactions

Like a human teacher the INTUITEL system should always be able to react to the learner. At best, the recommendation process is always based on sound considerations of all input information. But sometimes even a real teacher is not able to find the best recommendation to help the learner. Possible reasons for that could be the lack of some specific knowledge or the deduction problem is too complex to be solved in an appropriate time. In such cases teachers must rely on their intuition. Their "reflexes" kick in and the teachers give nearly-as-good answers as recommendations.

Ordinarily, the more experience a teacher has, the better the recommendations fit the learners' situation and needs. INTUITEL aims to do the same in order to provide for the best possible guidance. It thus needs a mechanism which decides when to bypass the reasoning process and when to trigger a direct reaction. Therefore, the LPM contains a "reflex module". Like in the real-life scenario, the reflex module acts when it is foreseeable that a response is not going to be created in time or when it is obvious that a certain action has to take place.

Because a recommendation cannot necessarily be created in all cases, the reflex module triggers a message that either informs the learner that his recommendation is pending (e.g., "I am still thinking about it, just one moment please.") or enforces the creation of a specific question that the learner should answer (e.g., "Which Learning Pathway do you want to choose? Select one of the following"). Another example would be to create a welcome message, if the learner just started a new session. For such a message, no computationally expensive reasoning has to take place.

3.3.4 LPM Input

This section defines and explains the different input types for the LPM. It firstly describes the learner dependent set of data. Further, the concept of Learning Pathways and their realization in the Pedagogical Ontology is outlined. Afterwards, the concept of the Cognitive Models is described.

3.3.5 Learner Input

Learner input basically describes all personalized learner information INTUITEL collects from and via the LMS and which can be used as a source for the recommendation creation process. This not only includes learner scores in terms of grades or results of tests, but also additional data which can be collected by INTUITEL. There is a multitude of different types or kinds of learner scores that provide diverse information about the learner from a learning habit, a learning progress, and a situational perspective. Depending on the possibilities of the LMS and the available information, some of it might also be collected by directly asking the learner. Examples of the available learner data are, apart from the rather obvious question how good the learner is in terms of grades, amongst others:

- What and when did the learner access content?
- In which order did the learner access learning content?
- How long was the learner working on certain a Learning Object?

- Which kind of device is the learner currently using?
- How good is the learners Internet connection?
- Is the learner currently in a noisy environment?
- Other possible raw data items may indicate emotional or stress measurements involving bodily interfaces to the learner.

In short, learner input describes all information on all aspects regarding how and what a learner currently and also previously has learned in the eLearning system across all courses. For a detailed list and description which data can be accessed through the respective INTUITEL interfaces, see the INTUITEL data model in Section 3.5. Examples could be age, name, gender etc.

3.3.6 Pedagogical Input

The pedagogical input for the LPM consists of two parts. On the one hand, there is the terminology specified in the Pedagogical Ontology. It provides a comprehensive set of entities for the modeling of learning material in INTUITEL. The structuring of learning material, independent of its actual LMS realization, allows a semantically rich description of e-learning courses. The central elements, namely CCs for abstract topic-based structuring of courses and KOs for the description of the actual content, allow the LPM to understand the meaning of the Learning Objects in the course. However, what is more important for the pedagogical input is the second aspect, the availability of Learning Pathways. In order to guide learners through an e-learning course, INTUITEL needs a "map" describing how to find a suitable route. This basically is what Learning Pathways (LP) provide for the back end. Modeling Learning Pathways with them, the LPM and the INTUITEL Engine are able to deduce a didactically reasonable route through the learning material.

3.3.7 Domain Input

The previously described pedagogical input alone is as itself insufficient for the back end, since it only provides the technical and didactic foundations. The description of the actual learning content is missing so far. In order to include this, the LPM needs the description of the knowledge domain from the lecturer. This is provided through the Cognitive Model (CM) which is created by a domain specialist. The CM outlines the curriculum of a certain domain of knowledge. This OWL-based description specifies which CCs are available in a specific Knowledge Domain (KD). Based on this a course in a LMS is connected by completing the course specification with the semantically rich description of the KOs. The thereby defined meta-data contains detailed information about the learning material itself (e.g., the contained media or what type of knowledge they represent).

To summarize, the domain input subsumes all information which the LPM and INTUITEL Engine need to understand the internal coherences in an eLearning course.

This does not mean that INTUITEL understands the content itself, i.e., what the learner is trying to learn, but is able to react on the semantic data as mentioned before.

A figurative example would be the simplified e-learning course "Math for absolute beginners" (see Figure 3.3). Consisting of a number of pages introducing to the basics of what numbers are and the basic arithmetic operations addition and subtraction, it depicts a very brief example of domain input in INTUITEL. On the first level, there is the start point of the course itself, the Knowledge Domain (KD). Branching from there, the individual CC are connected via macro LPs and allow navigating between them in a didactically reasonable way. The KOs is attached to the CCs and the micro LPs specify how a learner should best range between them on basis of the KO meta-data. With this information, the back end can deduce an optimal route for each individual learner, based on how many different LPs are available for that specific course.

Although this is not categorized as domain input, it should be noted that the individual Learning Pathways between the elements are firstly possible because the domain input provides a description of the content between which a route is possible. INTUITEL can thereby rely on the different macro LPs that the tutor has added into the Cognitive Model and also the micro LPs that are available due to their definition in the Pedagogical Ontology.

3.3.8 Set-based Rating of Learning Objects

Due to the usage of OWL-reasoning in INTUITEL, the approach of how recommendations are created is different to how other software solution approach that task. One of the main features of OWL-reasoners is the identification of elements in a set. This is a very useful functionality for INTUITEL, because creating a KO recommendation can be interpreted as the task of finding the elements of the set that contains the optimal KOs for the learner. Thus, the fundamental question of the back end is: What are the defining criteria of this set and how to evaluate them? Building on the previous introductions on what input is available for the LPM, it can be registered that there basically are three influential factors concerning the suitability of a KO:

- the learners macro Learning Pathway
- the learners micro Learning Pathway
- the situational dependent information of the learner

The first two points are relatively unproblematic, because this basically only requires data that is already available in the present system. This is at first the data which specifies the Learning Objects that are part of the course. Secondly, the history of the learner is needed to filter those LOs that are not available for the recommendation (e.g., because they are already finished). Thirdly, the learners personal Learning Pathway is required, to find the LOs that are next in the sequence. Given that this information is handed over to the Engine in a suitable way, the LOs that are relevant in context of the learners LP can be identified. This is possible, because the present problem

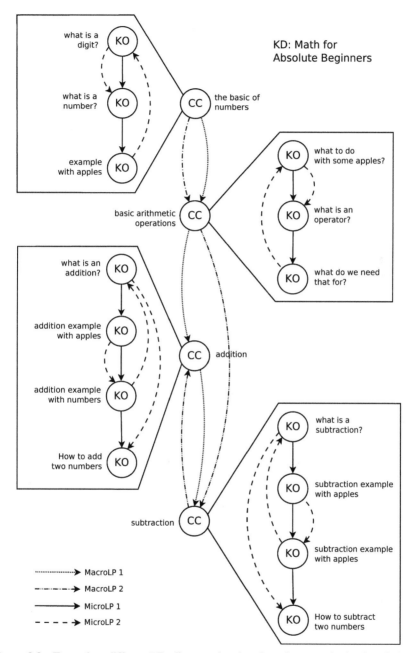

Figure 3.3 Exemplary different LPs diagram showing the coherences in the domain input, including different LPs.

can be described on basis of sets, allowing the Engine to iteratively reduce the number of LOs. In order to understand this procedure, it is advisable to illustrate this step-by-step.

As the first step, it should be clear that a LO is an umbrella term that combines CCs and KOs (for the sake of simplicity, Knowledge Domains (KDs), which are by definition also LOs, are excluded here). Each LO is thus either a CC or a KO. So, when starting with a set that contains all LOs of a certain course, it is possible to segment it into the set of CCs of that course and into the set of KOs of that course. These two sets can further be segmented into sets that differentiate LOs depending on whether they are rated as unseen, unfinished or already completed (Figure 3.4).

Independent of these sets is it possible to subdivide the set of available LOs of a course regarding their distance to the current Learning Pathway positions. For this, firstly the set of currently active KOs is identified which is represented by the small dark circles in Figure 3.4. This set of currently active KOs is either empty or contains exactly one item. It is empty if the learner has never worked on the course and has not yet accessed a KO in the current session. If the learner has already accessed a KO in the LMS, the set contains exactly this KO (i.e., the last one), which is rated as either unfinished or already complete.

Based on that, it is possible to identify the set of current CCs, which contains all CCs that has reference on the currently active KO. The set thus contains either one CC or multiple ones, because a KO can be part of more than one CC. The set of CCs that are next when following the macro LP is a logical consequence of the previous

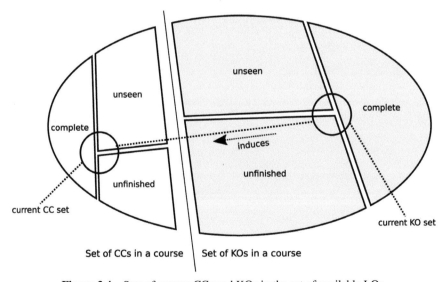

Figure 3.4 Sets of current CCs and KOs in the set of available LOs.

set. When it is known which CC(s) is/are currently active, the next CCs are those that are subsequent as described in the respective LP definition.

With these sets, the back end is able to identify the CCs that have recommendation relevance for the respective learner. This firstly is the intersection of the set of current CCs with the union of unfinished and unseen Ccs. If the set of CCs with recommendation relevance is empty, the course is either complete or there are no more CCs available in that particular macro LP. In the first case, a recommendation cannot be created. In the second case, a new macro LP needs to be selected or the process is finished too.

If at least one recommendation relevant CC has been identified, it is subsequently possible to create a list of KOs that have a CC-based relevance. This is the set of KOs that are attached to this/these particular CC(s). This is already a big reduction of the available LOs, but it can be reduced even further, when also including the micro LP information (Figure 3.5). This is the set that contains all KOs that are "next" as seen from the currently active KO and those KOs that are "next" regarding the "next" CCs. As previously, the term "next" is a matter of definition, which is based on the question what is reasonable to include here.

As the last step in the procedure, to identify those KOs that are recommendation relevant, the back end has to select those KOs that fulfill a list of requirements:

- KO is next or currently active
- KO is attached to a CC that is next or currently active
- KO must be unfinished or unseen (i.e., not already completed)

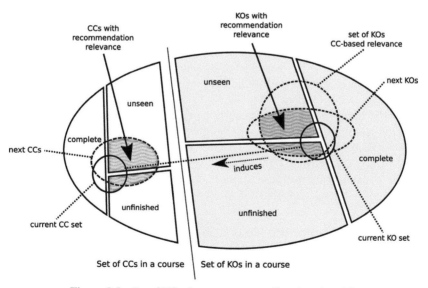

Figure 3.5 Set of KOs that are next regarding the micro LP.

Expressed on a set-basis, this is the intersection of the set of currently active KOs, with the set of KOs with CC-based relevance and the union of unfinished and unseen KOs. For an easier understanding, the diagram in Figure 3.5 clarifies these coherences graphically. If this respective set is empty, the same method as already used for an empty set of CCs with recommendation relevance is applied. Either has the micro and/or macro LP to be changed to include other LOs, or the process is finished, since there are no more KOs available to choose from.

When now coming back to the initial statement that there are three influential factors for the recommendation (macro LP, micro LP and situational dependent information), this so far only includes the first two aspects. In order to further personalize the recommendations the information that INTUITEL collects about the learner must be included in this procedure. For this, the back end specifies a method to include this information in a suitable way, which is implemented via the so called Didactic Factors that are explained in Section 3.2.

On the next level, these Didactic Factors are combined with properties of KOs to state their individual suitability for the learner. The procedure of connecting this information with the learning content is also expressible via the set-based approach as it has been applied to find the (LP-) relevant KOs. Therefore, for each Rating Factor, a set is specified that fulfills a particular rating rule. The task of the back end is then to find the elements that meet the respective conditions and to combine all this knowledge (Figure 3.6). This is done by calculating the intersection of the set of KOs with a general recommendation value (i.e., those elements that have been selected previously on basis of LPs) and those sets that express the optimum regarding the different Rating Factors.

3.3.9 Learning Progress and Learner Position

In INTUITEL the learner is being located in a multidimensional space. This section explains the intentions behind the Multidimensional Cognitive Space (MCS) and the associated Cognitive Content Space (CCS). Both are newly introduced concepts that the INTUITEL project uses to formalize the task of finding the position of the learner in the e-learning content. It has been designed to translate the basic educational conditions to a mathematical level, which is much better applicable on the domain of computer science. The presented geometrical representation of Learning Pathways and learning content in form of a multidimensional hypercube facilitates a better understanding of the technical realization of this task.

3.3.10 Determination of the Next KO

In the following it is outlined how to determine initial recommendations for a "next" KO to be processed by the learner, which are then modified by the Didactic Factors. Note, that throughout this section we are not dealing with Micro Learning Pathways

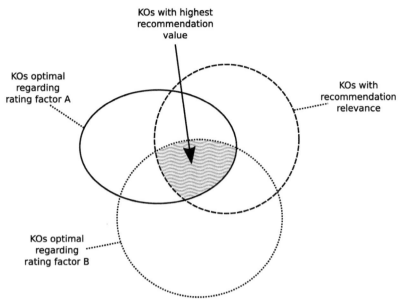

Figure 3.6 Set of KOs with the highest recommendation value.

or LOs that are considered alternatives, but only with Macro Learning Pathways and singular LP nodes. Consequently, for the purpose of this consideration, each KO may be considered as fundamentally different from all the others because it leads to a different knowledge strain. The basic paradigm of outcome-oriented constructivist learning is that it should lead to a certain learning goal. This goal might be quite complex, e.g., might require that

- a certain set of concepts is learned with a required precision
- a certain set of temporal, causal, or logical connections between these concepts is learned
- a certain set of methods and algorithms is learned which enables the learner to process new situations

Mathematically, this complex goal – which in fact is an ontology – may be expressed as a target position in an abstract space spanned by the concepts.

We term this the Multidimensional Cognitive Space MCS. Learning therefore may be considered a movement in this abstract space and learning should bring the learner closer to the target position. In modern human-centered teaching processes, teachers determine the current position of a learner by assessing the learners current knowledge and standing in regard to the topic and on a "meta"-level (e.g., how the learner learns). This happens either subconsciously by sensing that there are problems, delays or disturbances while the lessons are taking place, by evaluating the assignments of the learners (e.g., homework) or by having a direct conversations with them (e.g., questions during the lessons). These observations and feedback enables the teacher to come to conclusions regarding the learners' state of knowledge, habits, and characteristics to ultimately support the learner in the learning process.

INTUITEL provides means to carry out this tutorial guidance process also in technology enhanced learning, where no human teacher in traditional form is present and the learning so far may be seen as a self-directed process. A learning object recommendation, as it is generated by the INTUITEL system, therefore is a computer generated hint towards the self-directed learner on how the learner may come closer to the target position. Note that this recommendation is not entirely a global recommendation ("direct way to the target"), but may be locally deviant from the direct way to the target. Consider, for example, the possibility of a recommendation indicating that the learner should repeat a certain Learning Object: In this case it might correspond to a movement of the learner in the Multidimensional Cognitive Space (MCS) which for some time takes the learner further away from the final learning target position. The INTUITEL system therefore is handling the following tasks:

- Determination of the learner position in MCS
- Determination of a set of Didactic Factors influencing the learning process
- Application of the personal Learning Pathway on the course material
- Taking into account these Didactic Factors and the Learning Pathways, determination of the next step the learner has to take in order to ultimately come closer to the target position

The principles of the MCS will be reconsidered below, for now let us consider the Didactic Factors as they were explained in Section 3.2. Some of them are:

- Cognitive speed, which is the learning speed in the context of environment, personal attitude and current learner position is this a fast, a medium or a slow learner?
- Learning success is this a learner with superior, good, average or inferior results?
- Learning discipline does this learner follow suggestions easily, does the learner adhere to ascribed learning pathways?

For the moment, the INTUITEL consortium relies on a rather simple scale for each of these factors, e.g., "learning speed" is not measured in objective terms but in simple concepts of fast medium slow. This means, that a crucial module of the LPM is responsible to translate raw input data (from the LPM) into these Didactic Factors. Consider, for example the Didactic Factor of "learning speed". For each learner, the LMS delivers the actual learning time for each KO to the LPM. Furthermore, the LPM reads the metadata accompanying the learning content and knows the target learning time entered there by the cognitive engineer (henceforth called "estimated time"). The corresponding transformation rule for the LPM then reads, for example:

"If the learner needs more time than the estimated time to process a KO in at least 70 percent of all KOs, the learner will be assigned the value "slow" on the dimension of learning speed. If the learner needs less time than the estimated time to process a KO in at least 70 percent of the KOs, the learner will be assigned the value 'fast' on the dimension of learning speed. In any other case, the learner will be assigned the learning speed value "medium" (70-70-rule)."

However, suppose that the learner has already achieved a high level of knowledge, easy questions should then be answered faster. Therefore, one could think to shift from a 70-70-rule for learner speed determination to a 50-85-rule in this case. Or it may turn out that the learner is working in a high noise environment providing lots of distraction in this case one could revert to an 85-50-rule.

3.3.11 Cognitive Content Space

The Cognitive Content Space (CCS) is spanned by the Cognitive Model and the Semantic Learning Object Model (SLOM). SLOM is explained in more detail in Section 3.6. The Cognitive Model contains the pedagogical structure of a course; it is a concretion of the pedagogical ontology for a given domain of knowledge. The SLOM contains meta-data referring to the actual content. Both of these might contribute to the desired learning goal. For later development it is also possible that a more complex learning goal is defined in the Cognitive Model. For example, the cognitive engineer could specify, that in his Cognitive Model the desired learning goal puts more emphasis on practical knowledge than on theoretical knowledge. In such a case, the Cognitive Model would contain a transformation rule to determine the cognitive position from the raw input data.

3.3.12 Multidimensional Cognitive Space

The cognitive position of a learner is estimated by the amount the learner has learned from each of the KOs. Hence, the position for a course consisting of n KOs is determined by an n-dimensional vector within the Multidimensional Cognitive Space (MCS). The value of each vector element can be determined as follows:

- In a simple LMS one only knows that a KO has been processed. In this case we follow the strong optimistic learner assumption: Processing means learning, consequently the grade of progress of the respective KO jumps from 0 to 100 percent when a KO has been accessed.
- A more advanced LMS may tell INTUITEL which part of a KO has been processed by the learner. In this case we will follow the weak optimistic learner assumption: A partial processing of a KO means that the learner has learned the same percentage of this KO.

- A very advanced LMS will tell INTUITEL the result of a measurement, like, e.g., the percentage of points reached in a concluding test of this KO.

Let us now consider how a certain learning goal may be achieved as a sequence of cognitive positions. Let $P = (x_1, x_2, \ldots, xN)$ denote a vector representing the learners position in the Multidimensional Cognitive Space (MCS) with x_i representing the grade of progress of the ith KO.

The default learning goal states, that each KO has to be learned. The target position in the MCS, therefore, is a value of 100% = 1.0 for each component of this vector. The learner starts at position $Ps = (0, 0, \ldots, 0)$, his target position is $Pf = (1, 1, \ldots, 1)$.

Should there be a more complex learning goal (say, target learning profile) instead of all KOs being processed completely, the Cognitive Model defines a transformation, which for simplicity may be seen as a linear transformation (a matrix) reducing the dimensionality of the MCS from N to $M < N$. Therefore, without loss of generality, even in this case the same INTUITEL concepts and algorithms may be used as for the default learning goal.

If we assume a learner who completes each KO before moving on, the movement of this learner in MCS would always be along the edges of a N-dimensional hypercube: The learner has completed KOs 1, 2, 3, 4, $\ldots$, $k - 1$. He is currently working himself through the kth KO and still has to consider himself with KOs $k + 1$, $k + 2$, $\ldots$, N. Hence, his cognitive position would be

$$Lk = (1, \ 1, \ 1, ..., x_k, 0, \ 0, ..., 0)$$

Whereby x_k represents the KO at the current position. Here, without loss of generality, we have ordered the KOs in exactly the sequence as it is processed by the Learner, henceforth called User Learning Path (ULP). However, we might also assume a learner who is much less disciplined, and therefore does not complete any KO before moving on. Obviously, the sequence of cognitive positions in the MCS would be a curve inside (and not along the edges, but possibly across bounding surfaces) of the hypercube. In the extreme example, this learner would switch back and forth between the KOs in the model and finish each KO to the same degree. While this learner might nevertheless reach the final learning goal, his actual learning curve is a line close to the hyper-diagonal of the MCS. In order to achieve his final goal however, this learner would have to perform a very large number of switches between KOs, in order not to emphasize a particular KO.

For now let us assume that we have a disciplined learner who always moves along the edges of the knowledge hypercube. This amounts to a self-chosen sequence of KOs, each of them is processed (by assumption, learned) completely before moving on. After completing the k_{th} KO, the cognitive position would be

$$L_k = (1, 1, 1, \ldots, 1, 0, 0, \ldots, 0)$$

Obviously, if repetitions are excluded, this learner then has $N - k$ possible choices for his next KO and the direct cognitive distance to the target position is $\sqrt{N - k}$.

Each of those possible steps would reduce the direct cognitive distance by the same amount. Hence, none of the remaining KOs would be preferable from the viewpoint of reducing direct cognitive distance to the target state.

The Cognitive Model however may define certain Learning Pathways $LP_1, LP_2, \ldots, LP_S$. Since we reserved the standard numbering for the ULP, we have to consider each of these predefined Learning Pathways an ordered permutation of the numbers $1, 2, \ldots, N$. Since the current cognitive position after k learning steps may be arbitrarily close to Learning Pathway l but after a number of t steps we may not use simple path deformation rules to compare the actual current cognitive position to one that could be reached by this Learning Pathway. Therefore the LPM implements the following algorithm:

1. Determine the Learning Pathway LP_l from the Cognitive Model which runs closest (in distance d) to the current cognitive position L, and the closest corner point V (vertex) of P_l
2. Check each of the open choices $N - k$, if it would reduce the distance to V from d to $d' < d$ and assign to it the priority $d - d'$.
3. Check on the Learning Pathway P_l which the next KO would be after V, we assume that this would be KO no. v. If v is among the open choices $N - k$, add to the priority assignment of v the value $+1$ and terminate the loop. If not, choose the successor of v in P_l and check if it is among the open choice, add to its priority the value of 0.5, etc. Continue if either this loop terminates or if the end of P_l is reached.
4. Pass the information to the next decision stage.

The current cognitive position L is close to P_l and the next object recommended when P_l is followed by the one of the $N - k$ open choices which has the highest cumulated priority. If there is more than one next choice with the same priority, take a random selection among those. This algorithm ensures that a recommendation to the learner will, if the learner follows the recommendation, propagate through the MCS roughly in parallel to a certain Learning Pathway and at the same time towards this particular Learning Pathway. It is possible to relax this and present to the next higher decision stage a selection of alternative Learning Pathways and the corresponding suggestion for the next KO.

3.3.13 Example for Learner Positions

We now consider a system consisting of four different KOs A, B, C and D. The Multidimensional Cognitive Space then is a four-dimensional hypercube or tesseract. A three-dimensional projection of this hypercube is depicted in Figure 3.7.

The cognitive position of a learner is determined by the amount the learner has learned from each of these four KOs. Hence, the position is determined by a four-dimensional vector within the hypercube. The learner starts at position $Ps = (0, 0, 0, 0)$, which is the lower left of the inner cube of the above projection. His target position is

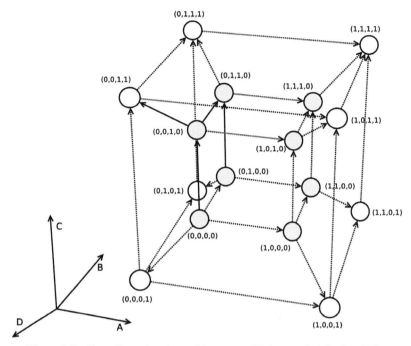

Figure 3.7   Four-dimensional cognitive space (4D hypercube) for four KOs.

$Pf = (1, 1, 1, 1)$, which corresponds to the upper right of the outer cube of the above projection. In principle $4! = 24$ distinct Learning Pathways exist and therefore, since the learner is free to choose, may be transgressed by the learner. Not all of these are didactically meaningful; hence let us assume that the KOs are grouped:

- The two pairs A, B, resp., C, D each constitute a thematic group.
- The pairs A, C, resp., B, D each constitute a chronological group. Such a grouping could arise, if
- A describes the contribution of Comenius to educational theory
- B describes the contribution of Habermas to educational theory
- C describes the contribution of Comenius to communicative didactics
- D describes the contribution of Habermas to communicative didactics

Consequently, two Learning pathways are defined in the Cognitive Model:

$$LP_h = (A, B, C, D) = (K_1, K_2, K_3, K_4) \text{ as } hierarchical \text{ LP}$$
$$LP_c = (A, C, B, D) = (K_1, K_3, K_2, K_4) \text{ as } chronological \text{ LP}$$

These two LPs differ only in one permutation of K_2, K_3. They are depicted by the thick arrows in the projection in Figure 3.8

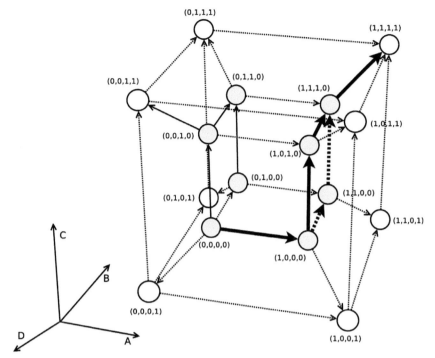

Figure 3.8 The two didactically meaningful learning pathways (hierarchical and chronological), shown as thick arrows.

Let us now follow a learner through this system. First of all let us assume a disciplined learner: The learner follows the initial advice and processes KO A, which puts the learner into cognitive position (1, 0, 0, 0). Obviously, the learner is still following both LPs defined in the Cognitive Model. If no other factor (like a pre-chosen LP) will lead the INTUITEL Engine to another conclusion, the learner will be offered two choices by the INTUITEL system:

> "If you want to follow the hierarchical LP, please process B. If you want to follow the chronological LP, please process C."

Conversely, we assume an undisciplined learner who, instead of following the initial advice to process A, has chosen to process D instead. This would put the learner after the first learning step into cognitive position (0, 0, 0, 1). Using the previously described algorithm, the LPM would determine that the learner is close to position (0, 0, 0, 0) which is on both well-defined LP. The learner would then get the advice to process A as the next KO, bringing the learner (if the advice is followed) into position (1, 0, 0, 1). There, the learner is close to position (1, 0, 0, 0), which is on both LP. Therefore, with equal priority (unless other factors come into play) the learner will be advised to process B or C. Suppose B is followed, this would put the learner into position $(1, 1, 0, 1)$ where the learner would finally get the advice to process C leading to the learning goal. The actual LP then is (D, A, B, C) where at least three KO are in the sequence which has been termed adequate by the cognitive engineer.

However, our learner might be so undisciplined that instead of following the first advice the learner takes a complete different route processing C after having done the first step of processing D. This would take the learner into position $(0, 0, 1, 1)$ and the closest of the well-defined LP's is the chronological LP with position $(1, 0, 1, 0)$. The learner would therefore get the advice "You are close to the chronological LP, please process B". If the learner follows this advice, the learner is led into position $(0, 1, 1, 1)$ where then the learner would get the final advice to process A and therefore the overall realized LP would be (D, C, B, A). In this realization, only the two KOs C and B are in the chronological sequence described in the Cognitive Model. We might also encounter a totally undisciplined learner, who also does not follow the second advice, e.g., processing first D, then C and then A. This would again bring the learner close to the chronological LP, and the learner would receive the final advice to process B. The actual LP then is (D, C, A, B) whereas in the previous scenario, only the two KOs A, B are in the chronological sequence described by the Cognitive Model.

3.3.14 Learner-State Ontology: The Output of the LPM

The output of the LPM consists of two parts. Firstly, there is the output of the actual task that the LPM carries out, the set of applicable Didactic Factors. These are defined in context of the Learning Model Ontology as elaborated in Sections 3.1 and 3.2. These Didactic Factors are essential for the individual learner-related rating of learning objects. The LPM prepares the input and particularly prepares the Didactic Factors that are passed to the Engine. It, for instance, checks the connectivity of the learner and ranks it, to state that the learner has a good or bad connection. This represents the learner-specific knowledge of the reasoner and allows the Engine to include this information in the deductive process.

However, the set of actually relevant Didactic Factors alone is not sufficient for the INTUITEL Engine. An OWL reasoner needs an ontology to perform its work and therefore, the LPM must create one that contains all relevant data, i.e., the second part of the LPM output. This basically is a merge of those meta-data that describe the respective course (including its Cognitive Model), the micro Learning Pathway information from the Pedagogical Ontology and the Learning Model Ontology, containing the machine-readable descriptions of Didactic Factors. To combine this information, the LPM creates the so-called Learner-StateOntology. This only temporarily valid collection depicts the current status of the learner in that course with respect to his or her learning history and current environment. Figure 3.9 illustrates the parts that are merged into the Learner State Ontology as the final output of the LPM.

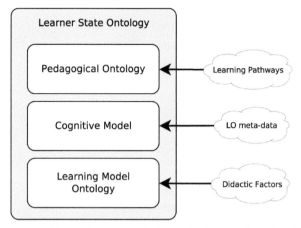

Figure 3.9 The final output of the LPM: the Learner State Ontology.

3.4 Software Architecture

Florian Heberle, Peter A. Henning and Kevin Fuchs

In the previous section we elaborated on the Learning Progress Model as an essential part of the INTUITEL concept. The concrete implementation of the LPM in form of a software component is a central part in the system architecture of INTUITEL. However, it is not the only one and it is embedded into a system landscape that comprises other components like the Learning Management System, a reasoning system and a recommendation component generating user-specific messages. The following section describes the overall system architecture proposed by the INTUITEL consortium.

The architecture of INTUITEL was designed with the focus on modularity and independence of technology. In particular it is insignificant which operating platforms or programming languages are chosen for implementation. Every single component may be implemented completely different from other components as long as the communication between them corresponds to the standard that is elaborated in Section 3.5.

The following describes the sub-components of the INTUITEL system. Note that they do not represent physical components in the first place. In stead they represent logical units that may but do not necessarily need to be reflected in separate software modules. Figure 3.10 shows an overview of the according components.

3.4.1 LMS Integration

The LMS integration component is in charge of tracking a learner within the LMS and informing the INTUITEL system about the learning objects the learner is accessing.

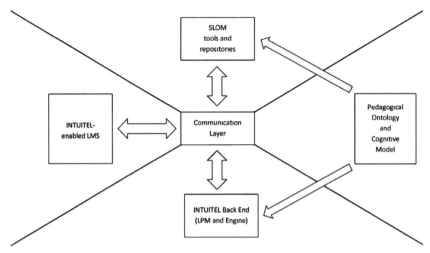

Figure 3.10   Software architecture.

Moreover the LMS has to provide a web service by which the INTUITEL system can access additional information about the learning environment, the progress of a learner as well as her or his individual characteristics.

Within the INTUITEL project the LMS integration was implemented in the form of plugins for four LMS: Moodle, Ilias, Exact and Crayons. Actually the INTUITEL standard only specifies the web service. It is up to the integrator how the necessary data are retrieved from the LMS and how they are transformed for the web service. In general, you do not even need a classical LMS. It is possible to enhance any learning environment as long as it fulfills the requirements of the web service specification. In Section 4.2, we discuss some example implementations of INTUITEL-compliant LMS enhancements. For example the web service of the LMS plugin must provide the following functionality:

3.4.1.1 Learner update

The INTUITEL back end has to be notified when a learner finishes a learning object or accesses a new one. For this purpose, the LMS plugin must implement a "learner update" mechanism. The LMS may behave in a active or passive way. For this reason, INTUITEL specifies update messages in a push or pull manner. The push scenario means that the LMS sends learner updates in a pro-active way to the INTUITEL back end. In pull mode, the LMS is polled by the back end for updates.

3.4.1.2 User Score Extraction (USE)

The User Score Extraction provides data from the LMS that are related to two aspects of an individual learner:

1. Performance data consist of information about a learner's learning behavior and progress in the past. This includes the completion status of learning objects a learner has attended to as well as specific scores. This data is needed to create personalized learning recommendations based on the learner's learning history and his current cognitive position respectively.

2. Environmental data provide information about the learning environment and the technical conditions under which a learner is working. This becomes especially important in the case of mobile learning which may involve permanently changing noise levels in the surroundings of the learner or temporarily low bandwidth due to mobile network connections. Moreover, environmental data also includes information about a learner's personal characteristics like her cultural background, country of origin, gender, age, physical or mental abilities and disabilities. Consequently such factors have an impact on the subset of recommendable learning objects. All this information may be contained in the database of the LMS or it may be actively prompted from the user via interactive message dialogues.

In concrete, the USE interface provides the data that is needed for the computation of Didactic Dactors (see Section 3.2).

3.4.1.3 Learning Object Recommender (LORE)

The result of the recommendation process taking place in the INTUITEL back end is a set of learning objects a learner is advised to access next. These learning objects are ranked by their levels of suitability according to the learner's learning history as well as her environmental and personal characteristics. The LORE module is responsible for the reception of these recommendations. The LMS integrator can decide freely how the recommendations influence the learner's experience within the user interface of the LMS. The recommendations may be presented as additional information in a pop-up window or an embedded message box. But it may also simply effect the order and priority respectively in which learning objects are listed in the LMS without the user even taking notice of the activities of the INTUITEL system. In Section 4.2 we discuss several examples of LMS integrations.

3.4.1.4 Tutorial Guidance (TUG)

The TUG web service of the LMS is in charge of receiving messages from the INTUITEL system and presenting it to the learner. Like in the case of LORE, TUG messages are the final output of the reasoning process. TUG messages may contain textual learning recommendations processed by a natural language unit that resides within the Recommendation Rewriter Component in the back end. TUG messages are not restricted to plain texts. They may also initiate an interactive dialogue between the learner and the INTUITEL system in order to gather more detailed information for the recommendation process.

For example, as explained before, the User Score Extraction module (USE) may need information about a learner's country of origin or her learning environment. In case that information cannot be gathered from the LMS it can be prompted from the learner directly by sending a respective TUG dialogue. This means that the USE interface can be emulated by the TUG module and that the according implementation of the USE module is not mandatory. The same is true for the previously described LORE interface, which can also be emulated by the TUG module. Therefore, the TUG interface is the minimum requirement that must be implemented by an INTUITEL-enabled LMS while USE and LORE are optional. This allows the implementation of very light-weight LMS plugins. Figure 3.11 illustrates the triggering of a single learner update and the consecutive message sequence between the LMS and the INTUITEL back end.

3.4.2 SLOM Repository

The SLOM Repository contains the Cognitive Models. They are based on the Pedagogical Ontology that we explain in Section 3.1 and describe didactic and pedagogical interrelations between learning objects in a form that can be read and

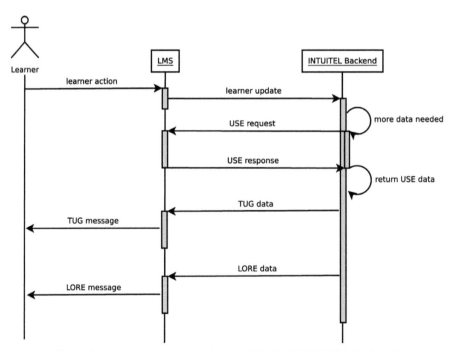

Figure 3.11   Message sequence between LMS and INTUITEL back end.

processed by a computer. In the particular case of INTUITEL it is the Reasoning Engine component that deduces on these ontologies. Also the LPM which is elaborated in Section 3.3 and further described below processes these Cognitive Models.

SLOM itself is basically a container format describing how cognitive models are represented by files. For a more detailed description of SLOM see Section 3.6. The SLOM repository manages these files and provides them for the other components. In order to create the SLOM files, content creators do not depend on specific software tools. Basically these files can even be created by hand with a simple text editor as they contain OWL ontologies in the form of XML statements. However, for larger contents this may not be feasible. Therefore, for the prototype, we have also implemented an editor with which arbitrary learning material can be imported from an LMS, annotated with the respective metadata and exported into SLOM files. Since all this is based on standard OWL, creators may also use any other OWL-compliant editor. While using standard OWL editors require knowledge in the field of OWL, our INTUITEL editor does not expect the creator to have such particular skills. The creator can create interrelations between learning objects by just drawing them in a drag-and-drop environment on the screen.

3.4.3 INTUITEL Backend

The INTUITEL back end comprises four sub-components which are the Learning Progress Model (LPM), the Query Builder, the Reasoning Engine and the Recommendation Rewriter. In the subsequent sections each of these components is explained.

3.4.3.1 LPM

The Learning Progress Model, as described in Section 3.3 plays a central role in the entire INTUITEL system and is implemented in the form of a respective back end component. It is responsible for all tasks that are needed to retrieve, analyze and transform the entire data set that is needed for the recommendation process. Any action that takes place in a learning environment is first processed by the LPM before passing any other component of the INTUITEL back end. In the following we explain the tasks which the LPM is in charge of.

Tracking of user actions In order to create any recommendations at all, the fundamental first step is to track a user's actions within the LMS. The LPM yields this information from the INTUITEL-enabled LMS via "learner update" messages. For this purpose the LPM can be configured due to multiple "communication styles". Depending on the respective LMS plugin, the required information is either sent actively from the LMS to the LPM each time a learner navigates to a learning object or performs any other relevant action like passing a test. Another method is having the LPM actively poll the LMS while the LMS itself remains in a reactive mode. Furthermore the LPM retrieves performance data from the LMS via the USE interface.

This data includes information about how well the learner has performed on particular KOs. Also using the USE interface, The LPM asks the LMS for additional user-specific information that are useful for the refinement of the recommendation process. This may be environmental data informing about the technical device the learner is working with, information about the surroundings of the learning environment or individual aspects like disabilities, cultural background or the gender of the learner.

Calculation of Didactic Factors The performance data; the environmental data and the information about a learner's individual characteristics form the combined input for the creation of Didactic Factors. Didactic Factors have an important impact on the recommendation process. Briefly recapitulating the elaborations in Section 3.2, the calculation of didactic factors consists of retrieving arbitrary non-nominal data from the LMS and transforming this data into a nominal representation of an entity within the ontologies that constitute the cognitive model. This way, the information is integrated into the ontologies and becomes processable for the reasoning engine.

Determining the learner's cognitive position The cognitive model, as derived from the pedagogical ontology describes the cognitive space in which the learner may move. Within this space the history of learning objects he has attended to combined with respective progress and achievements defines a vector we call the cognitive position of the learner. A key task of the LPM is to determine that cognitive position. Knowing the learner's current cognitive position, the INTUITEL engine is capable of comparing this position to The Learning Pathways that have been predefined within the Pedagogical Ontology by a didactic expert. By this a Learning Pathway can be determined that is closest to the current cognitive position.

Preparing data for the reasoning process Having collected all necessary data the LPM transforms them into a format that can be understood and processed by those components that are in charge of the reasoning process. For this purpose, all the data is packed into an ontology compatible to the pedagogical ontology and passed to the next stage which is the Query Builder.

3.4.3.2 Query builder
After all necessary data has been retrieved analyzed and transformed by the LPM, the Query Builder is the first entry point to the actual reasoning process. The main task of the Query Builder is first building a query for the next stage the Reasoning Engine and second optimizing that query with respect to specific requirements of the reasoner.

3.4.3.3 Reasoning engine
In the preceding, we described the LPM and how it creates an ontology firstly from the pedagogic ontology and secondly from the calculation of didactic factors. In that sense the main purpose of the LPM is to merge two domains of meta-knowledge

into one ontology: knowledge about the didactic and pedagogical concept underlying a course and knowledge about a learner's learning history as well as individual characteristics.

Being first processed by the Query Builder, which output ontology forms the input for the reasoning engine. In principle INTUITEL was designed to not only run one reasoning engine but multiple instances in parallel due to scalability. By default, INTUITEL provides one basic reasoning engine to cover the most common scenarios. In case multiple instances of reasoning engines are used, the architecture of INTUITEL allows using a reasoning broker as a preliminary authority before the reasoning engine. That broker is responsible for delegating different queries it receives from the Query Builder to the respective reasoning instance.

The reasoning process itself has to be understood in the following terms. The reasoner reads the ontology originally created by the LPM. Deducing from this ontology a reduction process on all available learning objects is performed. Within this reduction process, the rules and interrelations of learning objects are applied. At the end a subset of those learning objects remains that satisfies the rules encoded by the ontology.

3.4.3.4 Recommendation rewriter

Once the reasoning output has finished its task, it passes the result to the Recommendation Rewriter. As its name tells it "rewrites" the recommendation created by the reasoner into a human-friendly format. The result of this transformation are the ones previously explained TUG and LORE messages that are sent to the LMS plugin. The LORE messages contain a ranked list of suitable learning objects according to the reduction process of the reasoner. The TUG messages represent text messages that are generated by the help of a special natural language unit.

3.4.4 Data Exchange

The architecture of INTUITEL is formed by a modular design that requires only a loose coupling of the single components. The basis of the architectural design is a clear definition of web services that represent end points for the communication between the previously described components. Provided that these web services are implemented correctly, it is completely irrelevant which technologies are used to implement the single components. As long as the web services facilitate the correct data flow the functionality of the system is guaranteed. The way the INTUITEL system was implemented by the authors is therefore just one possible solution of a prototype and can be considered as a recommendation that may be modified according to particular requirements. Moreover, an INTUITEL implementation may be distributed running each component on its own physical platform or it may be completely or even partially integrated into one physical system. Altogether, this ensures a maximum of interoperability.

In order to decouple the single components from each other a fifth component is specified by the INTUITEL recommendations which are the Communication Layer. The Communication Layer is in charge of receiving and routing messages from components to the respective web services of the other components. Hereby the communication Layer hides the single components from one another and each component only has to address the Communication Layer as a single end point. A typical message sequence to generate a learning recommendation that is triggered by a learner accessing a new learning object from within the LMS is sketched below.

1. By accessing a new learning object, the learner triggers the LMS to send an update message to the Communication Layer (learner update) The Communication Layer routes this message to the component it considers being responsible which – in this case – is the LPM.
2. Through the Communication Layer the LPM contacts the SLOM repository and asks for appropriate SLOM data that are associated with the learning object the learner has just accessed. If needed, the LPM requests additional information from the LMS which includes data about the learning environment, performance data on the learner's progress so far and data concerning individual characteristics and preferences of the learner. This data is then processed resulting in a query that is passed to the next authority: the Query Builder.
3. The Query Builder takes the data it receives from the LPM and prepares an according query for the next stage – the Reasoning Engine.
4. The Reasoning Engine performs the reasoning process on the data it has received from the Query Builder and sends the result to the Recommendation Rewriter.
5. The Recommendation Rewriter is the last authority in the recommendation process. Taking the data it receives from the Reasoning Engine, the Recommendation Rewriter uses a natural language unit to create humanlike recommendation messages including ranked lists of related learning objects fitting most the current cognitive position of the learner.
6. Finally the Recommendation Rewriter sends LORE and TUG messages to the LMS.

Note that the recommendation process is triggered every time a learner accesses a new learning object. This is shown by Figure 3.12 which illustrates the above sketched message sequence. In the next section the data model of these messages is explained in more detail.

3.5 Data Model

Peter A. Henning and Florian Heberle

In the previous section we presented an overview of the INTUITEL architecture. We regard this as a general instruction on how to implement an INTUITEL system without any respect or limitations to technology and concrete software implementation.

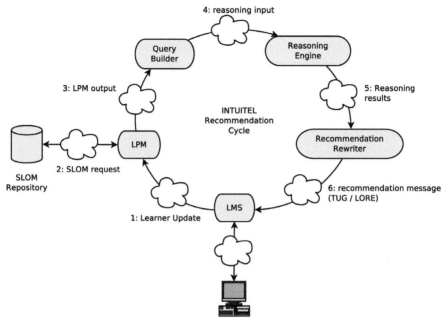

Figure 3.12 INTUITEL recommendation cycle.

The only requirement we claim is the way that the single components communicate with each other. In the following we give a clear definition of the communication interfaces. Any implementation of the INTUITEL system will work as long as these interfaces are implemented correctly.

The communication between the LMS and INTUITEL is implemented as a RESTful web service on both sides. REST (Representational State Transfer) is a series of design guidelines for web services, hence not a strict way to express communication paradigms. Table 3.3 exhibits a list of URLs that is used in this web service. It is

Table 3.3 INTUITEL end points

URL	Description
/lmsprofile	get some basic information about the LMS
/learners	Learner Update: provides information about learner actions
/login	used for annotation purposes: login as a course administrator
/mapping	used for annotation purposes: get a list of available learning objects
/TUG	Tutorial Guidance message
/LORE	Learning Objects Recommendation message
/USE/performance	get performance data for a specific learner
/USE/environment	get environmental data for a specific user

advised to run the communication over a proxy component – the Communication Layer as elaborated in Sections 3.4 and 4.1.

In the sequel we illustrate the data that is exchanged over these end points in more detail. Note that the following elaborations are supposed to give the reader a basic understanding of how the data exchange works. For a more detailed insight – especially if you intend to implement an INTUITEL system on your own – please refer to the technical documentation in the respective Deliverables of the INTUITEL Consortium [27, 28].

Initialization

The INTUITEL system is configured to interact with a certain LMS. When it is started, it will ask the LMS for its capabilities by a message which has the XML format:

```
<INTUITEL>
    <LmsProfile mId="some uuid"/>
</INTUITEL>
```

The parameter *mId* is a globally unique message ID. The LMS will respond with XML formatted data describing its capabilities and settings. The response contains the following data:

- A globally unique message ID, abbreviated as mId. This is the same message ID as in the request.
- A list of attribute-value pairs that provide information about the LMS. for detailed information see Table 3.4.

It will be returned as XML document according to the following template:

```
<INTUITEL>
    <LmsProfile mid="some uuid">
        <Data name="some name" value="some value"/>
        <!– – Repeated if necessary – –>
    </LmsProfile >
</INTUITEL>
```

Learner Update

After initialization, the INTUITEL system will react on learner navigation events. This means that INTUITEL will create recommendations (LORE and TUG messages) for learners, when they open a LO. To notify INTUITEL of such a LO transition, the LMS integrators can chose between three different methods. This allows them to select the most suitable one in respect to the technical possibilities and guidelines of their respective LMS.

Table 3.4 Attribute value pairs for the initialization response

Data Item Name	Description
lmsName	Installation name of LMS instance, e.g. "HS Karlsruhe"
lmsType	Type of LMS instance, for example "clix", "crayons", "exact", "ilias", "moodle", "other"
lmsId	Unique ID that allows identifying this particular LMS instance, like geolocation of LMS or registration number of LMS instance (may be left empty).
lmsMediaLevel	Level of multimedia support as a collation of characters: v = video, a = audio
comStyle	Specifies the method used for learner updates and how INTUITEL sends LORE and TUG messages: 0 = Pull, 1 = Push, 2 = Push with learner polling (see Section 3.5)

Pull Update

The objective of this scenario is to support LMSs that do not permit the usage of technologies that allow refreshing the LORE and TUG windows in the learners browser asynchronously (e.g., in a LMS that prohibits the usage of JavaScript). The LMS can thus include all necessary data in just one response to the learner. The learner update message sent by the LMS to INTUITEL contains XML-data in the following format:

```
<INTUITEL>
    <Learner mId="uuid" uId="user ID" loId="LO ID" time="access time"/>
</INTUITEL>
```

The Learner-element contains four attributes with the following meaning:

- A globally unique message ID, abbreviated as mID.
- The user ID which identifies the learner in the LMS, abbreviated as uId. This uId is provided by the concrete LMS installation and may vary according to organizational policy.
- The LMS specific Learning Object identifier, abbreviated as loId. This specifies the LO the learner just opened as new LO to work on.
- The point in time the learner accessed the LO in data type long, abbreviated as time.

INTUITEL's response to the LMS contains XML-data as described in the following template:

```
<INTUITEL>
    <Learner mId="some uuid">
    <Lore uId="some user ID" mId="some uuid"> . . . </Lore>
    <Tug uId="some user ID" mId="some uuid"> . . . </Tug>
</INTUITEL>
```

3.5.1 Push Update

In the push-scenario, the LMS sends a learner update message to INTUITEL when a learner opens a new LO. As in the pull use case, the INTUITEL back end reacts and creates a recommendation. However, the LMS will not get a direct response to the learner update message. It is just a notification for INTUITEL, triggering the reasoning process. The moment LORE or TUG data is available for a learner, the INTUITEL system sends individual messages to the LMS, which can then be forwarded to the learner.

This scenario is designed for LMSs that want to decouple the standard LO processing from INTUITEL. By using this approach, the LORE and TUG data can be loaded asynchronously in the learner's browser (e.g., by using JavaScript and AJAX) after receiving them from the INTUITEL back end. So, at the point in time when the TUG and LORE data is presented to the learner, he or she already started viewing the actual LO content.

The learner update message sent to INTUITEL contains XML-data in the format as specified in the pull scenario above. In order to afterwards assign the responses to the correct learner update message in the LMS, its respective mId is implemented in the LORE and TUG responses as rId (request ID). This ensures that each data item has a unique identifier and that the responses can be attributed to the messages which triggered their creation.

3.5.2 Push Update with Learner Polling

This method is similar to the previously described push approach. The important difference is that INTUITEL actively requests the learner update notifications via polling. Therefore, INTUITEL will ask the LMS in specific time intervals if a new learner update is available. If a LO transition is noticed in that process, the INTUITEL system will carry out the same workflow as in the standard push scenario.

Because the LMS does not have to send the learner updates to INTUITEL, it can act as an only receiving endpoint in the communication with INTUITEL. This allows LMSs that do not support active signaling, to still be able to use the functionalities of INTUITEL. However, this also includes that the learner update messages are adapted to fit to this use case. The messages sent to the LMS by INTUITEL contain XML-data with the following format:

```
<INTUITEL>
    <Learners mId="some uuid"/>
</INTUITEL>
```

As response to this message, the LMS will add a list of XML-elements describing the currently active learners. The mId of the learners-element is the same as in the request by INTUITEL. For each learner logged into the LMS this list will contain:

- A user ID which identifies the learner in the LMS, abbreviated as uId.
- A list which identifies the Learning Objects visited by this particular learner since the last poll. Each entry will also contain the LMS specific Learning Object id (loId) and the point in time the learner accessed the LO as long.

The data will be returned as XML document according to the following template:

```
<INTUITEL>
    <Learners mId="some uuid">
        <Learner uId="some user ID">
            <VisitedLo loId="LO ID" time="access time"/>
            <!-- Repeated if necessary -->
        </Learner>
        <!-- Repeated if necessary -->
    </Learners>
</INTUITEL>
```

As in the previously described push scenario, TUG and LORE messages contain the mId of this learner update as their rId. This allows the LMS to link the respective recommendation data to the learner update message, initially triggering the respective reasoning process.

3.5.3 Learning Objects Mapping and Inventory

The LMS provides a method to associate its internal content with the external INTUITEL metadata stored in the SLOM metadata repository (see Section 3.6 on the SLOM meta-data model). To this end, the INTUITEL Editor addresses the LMS by a message which has the following XML format:

```
<INTUITEL>
    <LoMapping mId="some uuid">
        <Data name="some name" value="some value"/>
        <!-- Repeated if necessary -->
            </LoMapping>
    <!-- Repeated if necessary -->
</INTUITEL>
```

Parameters are:

- A globally unique message ID, abbreviated as mId.
- Optionally: A collection of search strings in the form of attribute-value pairs which are used to narrow down the search results. For a detailed listing of the attribute-value pairs see Table 3.5.

Table 3.5   Attribute-value-pairs for the LO mapping

Data Item Name	Description
loName	Learning Object title or name (mandatory for response)
loId	Learning Object identifier as provided by the LMS (mandatory for response)
hasParent (if hierchical)	Learning Object identifier of parent LO as provided by the LMS (mandatory for response)
hasPrecedingSib (if sequential)	Learning Object identifier of preceding sibling LO as provided by the LMS (mandatory for response)
hasChild (if hierarchical)	Learning Object identifier of child LO as provided by the LMS, possibly multiple ones per LO
hasFollowingSib (if sequential)	Learning Object identifier of the following sibling LO as provided by the LMS
lang	language code of the LO
loType	Learning Object type as specified in the pedagogical ontology, e.g. "test"
learningTime	Typical time to work through this object
typicalAgeL	Typical learner age (lower boundary)
typicalAgeU	Typical learners age (upper boundary)
media	Comma-separated list of media types contained in the LO: "text", "image", "audio", "video", "interactive". To be extended if need arises
size	Size of the LO in terms of Kilobyte
getFullCourse	Attribute specifying that a full course mapping including all LO children and siblings shall be returned.

For each LO, the following data will be returned:

- The title or name (loName) and Learning Object ID (loId) of the particular object(s) found in the LMS and matching the search criteria.
- The position of the LO in the internal course structure:
  - *Hierarchical structure*: The LO's parent LO by specifying the attribute "hasParent" (if the LO is not on top level).
  - *Hierarchical structure*: The LO's child LO by specifying the attribute "hasChild" (if the LO is not on bottom level).
  - *Sequential structure*: The previous LO in the given sequence by specifying the attribute "hasPrecedingSib" (if the LO is not first in the LO sequence).
- Further meta-data obtainable from the LMS on this LO. For a detailed explanation consider Table 3.5.

The data will be returned as an XML document according to the following template:

```
<INTUITEL>
    <LoMapping mId="some uuid">
```

```
    <Data name="loName" value="some LO name"/>
    <Data name="loId" value="some LO ID"/>
    <Data name="hasParent" value="some LO ID"/>
    <!-- Repeated if necessary -->
    <!-- Order of data items is not relevant -->
</LoMapping>
<!-- Repeated if necessary -->
</INTUITEL>
```

3.5.4 Authentication

Please note that this authentication is not necessary for ordinary learners using an INTUITEL enabled LMS. It is for content creators who want to make use of the INTUITEL system to edit LO meta-data in the INTUITEL Editor.

To log a user into the INTUITEL system, the LMS will receive a message from INTUITEL containing the following parameters:

- A globally unique message ID (128 bit according to UUID standard), abbreviated as mId.
- A user ID which identifies the learner in the LMS, abbreviated as uId. This uId is provided by the concrete LMS installation and may vary according to organizational policy.
- A password abbreviated as Pass.

Each of these messages is an XML fragment enclosed in XML-elements named "Authentication" according to the following template:

```
<INTUITEL>
    <Authentication uId="some user ID" mId="some uuid">
        <Pass>some password</Pass>
    </Authentication>
</INTUITEL>
```

The LMS returns an XML document containing acceptance codes enclosed in XML-Elements as seen in the following template:

```
<INTUITEL>
    <Authentication uId="user ID" mId="uuid" status="OK or ERROR">
        <LoPerm loId="some LO ID"/>
        <!-- Repeated if necessary -->
    </Authentication>
</INTUITEL>
```

In each of these, the message ID and the user ID are identical to those contained in the original request. Contained in the Authentication-tag is a list of all courses that may be edited by this user as LoPerm-elements (Learning Object permissions). If the collection is empty, there are no courses available for editing for this particular user.

3.5.5 Tutorial Guidance – TUG

The TUG interface carries out a dialog between the learner and the INTUITEL system. The primary use case for the TUG interface occurs, when the INTUITEL system sends a message to a learner which contains the following structured information:

- A globally unique message ID, abbreviated as mId.
- A user ID which identifies the learner in the LMS, abbreviated as uId.
- A field containing a message type, which is a 4-digit number, abbreviated as MType. Please refer to Table 3.6 for a listing of the available types.
- A field containing the structured message data in XML format, abbreviated as MData.
- A globally unique request ID, abbreviated as rId. This rId specifies the ID of the message which triggered the creation of this TUG message.

Each of these messages is an XML fragment according to the following template:

```
<INTUITEL>
  <Tug uId="some user ID" mId="some uuid" [rId="some uuid"]>
    <MType>message type</MType>
    <MData>message data</MData>
  </Tug>
  <!-- Repeated if necessary -->
</INTUITEL>
```

Table 3.6   Minimum requirement for message types in the TUG interface

Mtype	Description
1	Simple message, not important. Represented as Text, if necessary with HTML formatting
2	Simple message, important. Represented as Text, if necessary with HTML formatting
3	Simple question, to be answered Yes/No. Represented as Text, if necessary with HTML formatting
4	Single choice question, to be answered with one out of n alternatives. Represented as Text, if necessary with HTML formatting; n different text pieces in structured writing
5	Multiple choice question, to be answered with any number out of n alternatives. Represented as Text, if necessary with HTML formatting; n different text pieces in structured writing

3.5.6 Immediate Response from the LMS

If the XML document containing one or more TUG messages has been received by the LMS, it will return an XML document containing one or more acceptance codes as seen in the following template:

```
<INTUITEL>
  <Tug uId="user ID" mId="uuid" retVal="OK, PAUSE or ERROR"/>
  <!– – Repeated if necessary – –>
</INTUITEL>
```

In each of these, the message ID and the user ID are identical to those contained in the original TUG message from INTUITEL. The acceptance code (retVal) is part of the fragment body and reads:

- OK: The learner addressed by the uId is known to the system and logged in. The TUG message has been accepted.
- PAUSE: The learner addressed by the uId is known to the system but not logged in. He may log in after pausing for some time according to his personal learning schedule. The TUG message has temporarily been rejected. The INTUITEL system will have to resend and possibly update it at a later point in time.
- ERROR: The learner addressed by the uId is not known to the system. The TUG message has been rejected.

3.5.7 Delayed Response from the Learner

If the state of the return message for a particular learner is "OK", the LMS presents on the screen of this particular learner a message window or a similar widget. The exact implementation is left to the individual LMS. This window shall be displaying a message or asking for learner action, e.g., asking the learner to fill in text, click boxes etc.

It cannot be guaranteed, that a particular learner will process a TUG message at all or within a certain response time. If a TUG message has been answered by the learner by pressing the "Submit" button, even after some arbitrary response time, this will be communicated to the INTUITEL system. Therefore, a message from the LMS to the INTUITEL system will be prepared according to the following template:

```
<INTUITEL>
  <Tug uId="some user ID" mId="some uuid">
    <Data name="some name" value="some value"/>
    <!– – Repeated if necessary – –>
  </Tug>
  <!– – Repeated if necessary – –>
</INTUITEL>
```

3.5.8 Learning Object Recommender – LORE

Recommendations are produced as part of the didactic reasoning process in the INTUITEL Engine and are sent to the LMS. The main use case for the LORE interface therefore occurs, when the INTUITEL system recommends to the learner a list of Learning Objects and their priority containing the following structured data:

- A globally unique message ID, abbreviated as mId.
- A user ID which identifies the learner in the LMS, abbreviated as uId.
- A list of elements containing the Learning Object identifier (abbreviated as loId) paired with the allocated priority. This priority is considered to be an integer value from the interval (0, 100) denoting the percentage of relevance.
- A globally unique request ID, abbreviated as rId. This rId specifies the ID of the message which triggered the creation of this TUG message.

Each of these messages is an XML body according to the following template:

```
<INTUITEL>
        <Lore uId="some user ID" mId="some uuid" [rId="some uuid"]>
        <LorePrio loId="LO id" value="a value between 1 and 100"/>
        <!-- Repeated if necessary -->
    </Lore>
</INTUITEL>
```

All other LOs in that particular course, which are not contained in the LORE element, indirectly get a priority value of zero. So, each time the LMS receives a new LORE message, the previous recommendation can be rated as outdated. If the XML document containing one or more LORE messages has been received by the LMS, it responds with an XML document as described in the following template:

```
<INTUITEL>
    <Lore uId="user ID" mId="uuid" retVal="OK, PAUSE or ERROR"/>
    <!-- Repeated if necessary -->
</INTUITEL>
```

In each of these LORE elements, the message ID and the user ID are identical to those contained in the LORE message from INTUITEL. The acceptance code (retVal) is part of the fragment body and reads:

- OK: The learner addressed by the uId is known to the system and logged in. The recommendation has been accepted.
- PAUSE: The learner addressed by the uId is known to the system but not logged in. He may log in after pausing for some time according to his personal learning schedule. In this case, no further action is taken because INTUITEL

recommendation should pause as well. The recommendation has temporarily been rejected. The INTUITEL system will have to resend and possibly update it at a later point in time.

- ERRROR: The learner addressed by the uId is not known to the system. The recommendation has been rejected.

3.5.9 User Score Extraction – USE

The USE interface provides the INTUITEL system with personal data from the LMS concerning a particular learner. Note that the USE interface has two important uses, because it may be used to get information about the learner's performance, as well as on the environment.

3.5.9.1 Primary use case: Performance data request

The main use case of the USE interface is to request information about a learner's performance according to specified LOs. This allows INTUITEL to include the learner's current skill level in the reasoning process to create recommendations with a much higher level of personalization.

To receive personal learning progress related data from the LMS, INTUITEL sends a USE request specifying the needed data. Therefore, the UsePerf-element contains the following structured data:

- A globally unique message ID, abbreviated as mId.
- A user ID which identifies the learner in the LMS, abbreviated as uId.
- Optionally: A list of elements (LoPerf) specifying Learning Object IDs, abbreviated as loId.

Each of these messages is an XML body according to the following template:

```
<INTUITEL>
    <UsePerf uId="some user ID" mId="some uuid">
        <LoPerf loId="some LO ID"/>
        <!- - Repeated if necessary - ->
    </UsePerf>
    <!- - Repeated if necessary - ->
</INTUITEL>
```

If the loId-attribute is left blank, the LMS shall return all available LO scores for that particular learner. This shortcut will be used to get all available learner scores in one request when learners start their very first learning session with INTUITEL. If the XML document containing one or more USE messages has been received by the LMS, it will return an XML document containing a body as seen in the following template.

```
<INTUITEL>
    <UsePerf uId="some user ID" mId="some uuid"/>
        <LoPerf loId="some LO ID">
            <Score type="score type" value="score value"/>
            <!-- Repeated if necessary -->
        </LoPerf>
        <!-- Repeated if necessary -->
    </UsePerf>
    <!-- Repeated if necessary -->
</INTUITEL>
```

In each of these, the message ID and the user ID are identical to those contained in the original request. Since there are multiple different types of scores, the contained Score-element further specifies the meaning of the respective value. Which type of score will be relevant for a LO varies depending on the respective LO type and LMS. Not every LMS collects the same amount of data. Consequently, INTUITEL does not require the LMS to fulfill certain requirements concerning these USE scores. A concrete INTUITEL implementation may for example provide the following score types:

- *grade*: The grade a learner achieved in a LO. Values are expected to be within [1, 6].
- *completion*: The completion status of the LO in terms of a percentage value, should be used in context on textual LOs.
- *seenPercentage*: The percentage of the LO which has been seen by the learner. It should be used in context of video and audio LOs.
- *accessed*: Boolean value specifying whether a learner has already accessed a LO (true) or not (false).

It is up to the developers to implement any other score types.

3.5.9.2 Secondary use case: Environmental data request

The second use case of the USE interface is to request information about learners themselves and about their current environmental situation. This allows INTUITEL to further personalize the TUG messages and the recommendation creation. Furthermore, this allows the back end to include the situational aspects (location, device, . . .) of the learner's current learning session in the reasoning process. The environmental USE request to the LMS contains the following structured query parameters:

- A globally unique message ID, abbreviated as mId.
- A user ID which identifies the learner in the LMS, abbreviated as uId.
- Optionally: A data item name specifying the requested data item (see Table 3.7).

Each of these messages is a XML body according to the following template:

Table 3.7 Parameters describing the learning environment of the learner

Data Item Name	Description
IName	Full name of the learner, which will be used in context of TUG-messages
IGender	Learner's gender
IAge	Learner's age
ICulture	Learner's culture. Values are "asian", "caucasian", "african"
IAttitude	Current emotional attitude of the learner. If natively available via LMS; e.g. analysis of facial expression
eNoiseLvl	Average noise level as measured by the microphone of the device
eTime	Current daytime of the learner
dType	Currently used type of device: "desktop", "laptop", "tablet", "smartlet", "phone"
dConType	Type of connection used to access the LMS: "wire", "lte", "hsdpa", "umts", "edge"
dConStab	Number specifying the stability of the internet connection, value between 0 and 100
dRes	Screen size of the currently used device (horizontal x vertical)
dBattery	Current battery state in percentage value

```
<INTUITEL>
    <UseEnv uId="some user ID" mId="some uuid">
        <Data name="some name"/>
        <!-- Repeated if necessary -->
    </UseEnv>
    <!-- Repeated if necessary -->
</INTUITEL>
```

If the XML document containing one or more USE messages has been received by the LMS, it will return an XML document containing a set of different data items, as described in the following template:

```
<INTUITEL>
    <UseEnv uId="user ID" mId="uuid" retVal="OK, PAUSE or ERROR">
        <Data name="some name" value="some value"/>
        <!-- Repeated if necessary -->
    </UseEnv>
    <!-- Repeated if necessary -->
</INTUITEL>
```

In each of these, the message ID and the user ID are identical to those contained in the original request. The attribute-value pairs in the "Data" tag correspond to the items listed in Table 3.7. The acceptance code is part of the fragment body, the retVal string reads:

- OK: the learner addressed by the uId is known to the system and logged in.
- PAUSE: the learner addressed by the uId is known to the system but not logged in. He may log in after pausing for some time according to his personal learning schedule. In this case, no further action is taken because INTUITEL recommendation should pause as well.
- ERROR: the learner addressed by the uId is not known to the system. The response is dependent on the retVal and the specified Data-element in the request.
- In case of PAUSE or ERROR, the "UseEnv" element is empty.
- In case of OK and if one or more explicit data items were contained in the request, only data for these items is returned.
- In case of OK and if no explicit data item was contained in the request, all currently available environmental data items are returned.

3.6 The Semantic Learning Object Model SLOM

Stefan Zander and Florian Heberle

In addition to the data model for the communication between the INTUITEL components, which we explained in the previous section, we now briefly introduce the Semantic Learning Object Model (SLOM). SLOM is a new meta-data model developed within the INTUITEL project that provides a semantic web-based knowledge representation framework for the interlinkage of pedagogical and domain-specific knowledge with concrete learning material. SLOM's objective is to complement existing and well-known eLearning formats such as Sharable Content Object Reference Model (SCORM) and IMS-Learning Design (IMS-LD) with semantic information. This allows INTUITEL-enabled Learning Management Systems to process learning material represented as SLOM in more intelligent and personalized way and adapt internal decision making processes to the personal preferences and learning style of learners. It further serves as facilitating data infrastructure for the utilization and integration of externally hosted data in INTUITEL-compliant learning material.

SLOM also provides a data format specification in which the INTUITEL system stores general information about courses. This is a necessary prerequisite for the computation of learning recommendations and personalized feedback for learners. The SLOM format is implemented as a direct extension of the Pedagogical Ontology and provides additional language primitives together with a prescribing schema that determines how course information needs to be described and annotated in order to be compliant to an INTUITEL system. SLOM as a metadata format encloses two different ontologies with varying granularity:

- the Cognitive Map (CM) and
- the Cognitive Content Map (CCM).

While the CM provides terms for the description of topics in a knowledge domain of a course, the CCM serves as description framework for the actual learning material for a given course. A CM is intended to be universally valid and thus can be reused across different courses pertaining to a given topic. CCMs, in contrast, are specific to a given course since it represents the actual learning content.

In its second function as a storage format, SLOM provides a container structure for the original learning material the CCM is based upon. The storage structure specification of SLOM prescribes the structure in which compliant learning material needs to be stored. This entails three main pillars of information that should be compiled into the CM/CCM from the original content format:

1. Topology: The topological pillar contains information about the elements that constitute concrete learning material plus their topological structure and coherence. In terms of an INTUITEL system, a SLOM content package contains definitions for CC and KOs of a given domain of knowledge.

2. Sequences: Sequences are represented as Learning Pathways (LPs) and define the interlinkage of KOs and CCs on different levels of granularity. They are a main pillar of the INTUITEL approach and provide course instructors and designers the opportunity to orchestrate learning content in different didactically meaningful ways. Macro Learning Pathways (MLPs) define the sequence through which learners should work through the CCs of a course. Micro Learning Pathways (mLPs) define sequences how learners should work through KOs – the actual contents of a course. The total amount of possible learning pathways is the result of a combination of Macro and Micro Learning Pathways. While only a manageable set of Learning pathways is modeled by a course instructor or designer, the total amount of didactically meaningful ways to proceed through a given course is the product of them. In technical terms, learning pathways are represented as directed acyclic graphs, otherwise they could not be processed by a reasoner. A detailed description of this was given in Sections 3.1 and 3.3.

3. Technical and Educational Information: The third pillar of information refers to both CCs and KOs likewise: It concerns concrete information about how the content of KOs is to be displayed in terms of technical conditions (e.g., recommended screen resolution, color depth of an image, etc.) as well as educational metadata reflecting pedagogical purpose and background. Such annotations could comprise information about appropriate difficulty levels (e.g., beginner, intermediate level, expert) as well as the knowledge type a KO contains (e.g., varying assignment types).

A combination of these three pieces of information with learner data allows the INTUITEL system to create recommendations for appropriate learning objects and to produce feedback messages in that process. The more information that can be provided on the course, the more information can be integrated in this process. For more detailed information – especially the technical implementation of SLOM – refer to Deliverable 4.1 [29] of the INTUITEL consortium.

4

Prototype Implementation

4.1 Back End

Kevin Fuchs

In Section 3.4 we defined the software architecture for a real implementation of the INTUOTEL system. This architecture is just a recommendation given by the INTUITEL consortium. However, as long as you follow the data model and the communication standards given in Section 3.5, you may design an architecture of your own.

Our recommended architecture comprises six logical components that cooperate in the form of a recommendation cycle as illustrated in Figure 4.1. As a seventh component, we recommend a communication layer to act as a proxy mediating the communication between the components. Through this way components are decoupled from each other. Every single component only communicates with the communication manager as if it was communicating with the respective component. This guarantees a high level of interoperability as each component may be replaced or modified at any time without any effect on the other parts of the system.

The system may or may not be implemented in a distributed way with each component being deployed as a single part. Your choice will depend on the question how scalable your system is supposed to be.

The INTUITEL consortium implemented a prototype back end that was partly distributed. The LPM, the Query Builder and the Reasoning Engine were integrated into one physical component but with each sub component having its own REST interface. The Recommendation Rewriter is a component of its own and as recommended above, we implemented a Communication Layer. The SLOM repository was implemented as a simple local directory on the LPM system. In this folder the SLOM data was stored in the form of XML files containing the OWL descriptions of the ontologies. The annotation of course material is done by simply uploading these files into that directory.

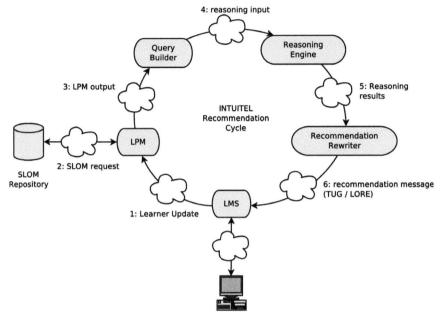

Figure 4.1 INTUITEL recommendation cycle.

For each component only the respective REST end point as shown in Figure 4.1 has to be configured. By using a Communication Layer as a mediator it is sufficient to configure only that Communication Layer as end point for each component. The Communication Layer is the only one that is aware of all other components. Also, the LMS is configured to the Communication Layer end point. This is especially convenient for the development of LMS plugins that then only have to consider that single end point. Its configuration contains routing tables for all message types in order to direct the communication from a specific source to its correct destination. As an example, Table 4.1 lists the routing table for all messages that are needed for the communication with the LMS plugin.

The prototype software can be downloaded from the INTUITEL website[1]. The LPM including the Query Builder and the Reasoning Engine is implemented in Java and can run on any platform. The Recommendation Rewriter is also a Java program. The Communication Layer is written in C# which means that you will need a Windows platform. However, using the Communication is optional although we recommend it. If you do not use the Communication

---

[1]http://www.intuitel.eu/resources/

Table 4.1 Routing table of the Communication Layer for LMS communication

Source	Message Type	Destination
LPM	LmsProfile request	LMS
LMS	LmsProfile response	LPM
LMS	LearnerUpdate request	LPM
LPM	LearnerUpdate response	LMS
LPM	UseEnv request	LMS
LMS	UseEnv response	LPM
LPM	UsePerf request	LMS
LMS	UsePerf response	LPM
Recommendation Rewriter	LORE request	LMS
LMS	LORE response	Recommendation Rewriter
Recommendation Rewriter	TUG request	LMS
LMS	TUG response	Recommendation Rewriter

Layer you will have to configure the end points for each component as shown in Figure 4.1.

4.2 LMS Plugins

In order to prove the concept of INTUITEL, we now introduce four exemplary implementations of LMS extensions. We provide extensions for Moodle, Ilias, eXact and IMC Learning Suite. These extensions provide the interfaces as defined by the data model in Section 3.5 to make INTUITEL communicate with the respective LMS.

Let us briefly remember the functionality which a LMS extension has to implement. First, a Learner Update message has to be implemented to inform INTUITEL about learner actions. Second, the LMS extension has to provide a USE interface by which INTUITEL can yield performance and environmental data about a learner. Third, the LMS extension must be capable of receiving and displaying TUG and LORE messages. Fourth, the LMS extension has to provide an interface for the INTUITEL Editor.

The definition of these interfaces is exactly the same for any LMS extension. The INTUITEL backend is indifferent to which LMS it connects to as long as the communication standards, which were defined in Section 3.5, are satisfied. Therefore, in the following, we will provide an altering view, focusing on a different aspect of the implementation for each extension. You will realize that the architecture and implementation of the extensions is very different. This is for two reasons: First, the implementation depends on the respective LMS. Second, except for the clearly defined interfaces that are needed for the communication between the extensions and INTUITEL, there are no restrictions with respect to the way of implementation or the technology that is used. The introduced implementations therefore include a variety of technologies like Java, JavaScript, C#, PHP, mySQL etc. Actually this emphasizes INTUITEL's claim to be a technology-independent approach.

4.2.1 Moodle

Elena Verdú, María J. Verdú, Luisa M. Regueras and Juan P. de Castro

Moodle is a pure-HTTP platform developed with accessibility and portability in mind. It has several extension mechanisms that can be used to add code to different stages of the cycle of life of page generation. The INTUITEL extension is implemented using the API for building Blocks. Blocks are a special type of plugins that render a box with contextual content. They are configurable to be embedded besides the main content of many pages. This is the mechanism of choice to implement the communication with the student and to trigger notifications of events to INTUITEL services from the LMS.

With respect to INTUITEL, there is not any communication channel to the student that can be started by the LMS; all information should be pulled by the student's browser. Hence, the Moodle adaptor relies on the persistence of the TUG and LORE messages in the INTUITEL server to function. This approach avoids dealing with messages timing and sequencing and takes advantage of the return channel of Learner Update requests for getting TUG and LORE content.

There is no problem to implement REST service points because every script in the PHP server acts as a potential REST service. Every access to these services is checked against a white-list of allowed remote clients to secure the integration with INTUITEL engines. This list is maintained by an administrator in the global settings of the Moodle platform. All parameters used in SQL queries are filtered using a Moodle API to avoid SQL injection attacks. This behavior fulfills the security requirements of INTUITEL system.

4.2.1.1 Implementation and architecture

Being implemented as a block of Moodle, the source code of INTUITEL interfaces must be placed in the folder "blocks" of Moodle within a folder called "intuitel": [MoodlePath]/blocks/intuitel. Figure 4.2 shows the main components of the architecture of the implementation of the interfaces for Moodle.

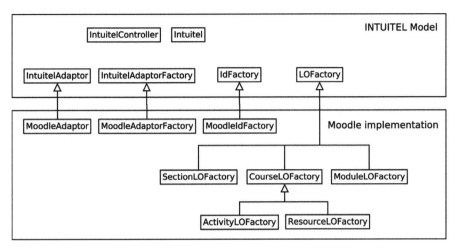

Figure 4.2   Architecture of the Moodle-plugin.

In order to allow the reusability of the model in future implementations or for other LMSs, the classes shown in the frame "Intuitel Model" in Figure 4.2, define the needed data structure, and define, and/or implement, all the operations that allow any LMS to operate with INTUITEL. The classes included in the "Moodle implementation" frame are specific to Moodle, specializing the methods to adapt the data structure and operations to Moodle.

As shown in Figure 4.2, MoodleAdaptor, MoodleAdaptorFactory and MoodleIDFactory are specializations of their respective parent classes needed to perform the defined operations in Moodle. The different descendants of LOFactory are also needed to create the specific LOs of Moodle. Next, some operations of the INTUITEL protocol are going to be detailed. Note that these operations have to be implemented for any LMS extension. Therefore, the following elaborations also apply for the other LMS extensions.

Lmsprofile: Specific capabilities of the Moodle instance are specified in the response to an LMSprofile request. The method to obtain the capabilities is implemented in any descendant of IntuitelAdaptor (i.e. class MoodleAdaptor). It simply returns an array of properties used to compose the response, which include lmsName, husType (fixed as "Moodie") husId (a configuration parameter) and comStyle (fixed as pull communication).

Leamer Update: In order to notify INTUITEL about LO transitions, each interaction of the student triggers an event notified to INTUITEL by means of a LeamerUpdate request that expects in return a TUG and a LORE message piggy-backed in the response (Figure 4.3), By default, only one event is

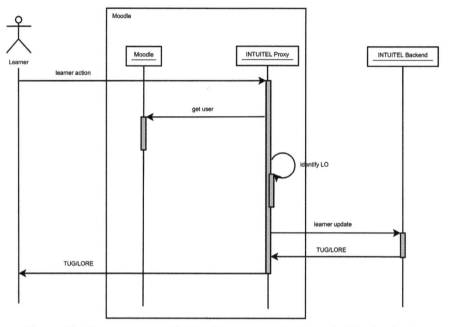

Figure 4.3   Message sequence for learning recommendations in the Moodle-plugin.

reported although the implementation allows also reporting a list of events from the event log, For every user, the system registers the last time when the visited LOs have been informed to INTUITEL.

LO Mapping and Inventory: The INTUITEL editor needs to retrieve the LOs in a course to allow the content creator to create the pedagogical modeL This service returns the structure of a Moodie course including sections and modules.

TUG and LORE: As there is no way to deliver any message to the user in response of a TUG or LORE request from the INTUITEL server, the only answers are: return an "error" response message if the user is unknown; or return a "pause" response message to tell INTUITEL that the user is not aware yet of the message but he is expected to visit a page in the future.

Effective TUG messages and LORE recommendations are received as responses from the INTUITEL server to the Learner Update requests sent by the LMS (Figure 4.3), therefore, they are processed each time the INTUITEL block is loaded in the browser (that is, each time the user accesses a new LO). The IntuitelProxy class is in charge of forwarding both the Learner Update requests from the block in Moodle to the INTUITEL server and the

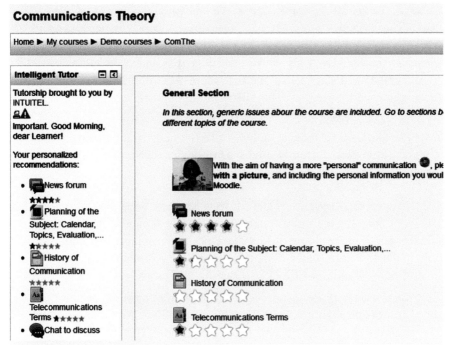

Figure 4.4 Recommendation messages in the Moodle GUI.

TUG requests when some user answers a TUG message displayed in the block. The implementation displays LORE recommendations within the block content. Besides, it modifies the DOM model of the page adding stars to the LO addressed (Figure 4.4).

USE: The USE Performance request has as response, for each user, a list of scores in each LO contained in every INTUITEL-enabled course the user is enrolled in. There are five LO score types, as defined in the data model (see Section 3.5):

- *Grade*: if the LO is configured as gradable (in Moodle activities can be configured as gradable or not by the teacher) and the student has received a grade, the score item will be included in the response for that LO.
- *Completion*: it will be provided in the response if the teacher has enabled the completion tracking for the activity in Moodle. The completion status will be 0% or 100% as completion tracking in Moodle does not implement partial completion tracking.

- *Accessed*: For all LOs this score type is provided in the response, specifying whether a learner has accessed a LO or not, according to the logs registered by Moodle.
- *SeenPercentage*: For LOs whose media type is video or audio, a 0% is indicated in the response for this score type if the object has not been accessed by the user, and 100% in the opposite case. There is no possibility in Moodle to track a partial percentage.

The response to a UseEnv request includes environmental data of one or more users: name of the user, currently used type of device, current daytime of the learner and the current location of the user.

4.2.1.2 How to configure INTUITEL in a Moodle site

INTUITEL is implemented as a block and therefore it can be installed in a Moodle site as any other plugin of Moodle. Once it is installed, it is important to configure a few different parameters (menu → Administration → Site administration → Plugins → Blocks → Intelligent Tutor):

- *Identification prefix for this Moodle instance*: This string is used as part of identifiers throughout the INTUITEL system. It is used to ensure that the ids are unique and relevant for this platform. The default value is taken from the unique identification of the Moodle installation, but this can be set to another meaningful value or to retain an old prefix in case the server has been moved and need to reuse existing INTUITEL activities and configurations.
- *List of IPs allowed sending INTUITEL events to this LMS*: this is a white-list of the hosts allowed to send messages and requests to this instance. Only IPs of trusted INTUITEL servers should be included here.
- *Service Point IP for using INTUITEL services*: location of the remote INTUITEL servers.

4.2.1.3 How to enable INTUITEL in a Moodle course

INTUITEL in Moodle is presented as a standard Moodle block with name "Intelligent Tutor" that can be added in any course. By default, the users with role "teacher", "editing teacher", or "manager" will be able to add the INTUITEL block in the course. With the edition option activated, the user can activate the INTUITEL interfaces for the course by adding the block "Intelligent Tutor" into the course.

Next step is to check the configuration of the block by clicking on the corresponding icon so that the configuration options are displayed (see Figure 4.5).

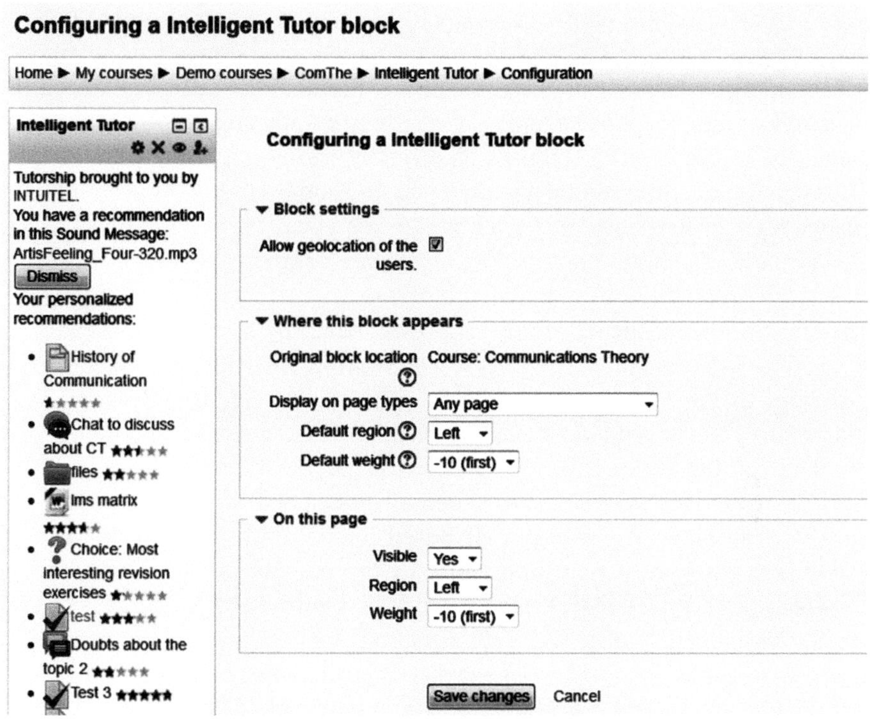

Configuring a Intelligent Tutor block

Home ▶ My courses ▶ Demo courses ▶ ComThe ▶ Intelligent Tutor ▶ Configuration

Intelligent Tutor ☐ ⬓	
✿ ✕ ◉ ♣	**Configuring a Intelligent Tutor block**

Tutorship brought to you by INTUITEL.
You have a recommendation in this Sound Message:
ArtisFeeling_Four-320.mp3
[Dismiss]

Your personalized recommendations:

- 📄History of Communication ★★★★★
- 💬Chat to discuss about CT ★★★★★
- 📁files ★★★★★
- 🎞ims matrix ★★★★★
- ❓Choice: Most interesting revision exercises ★★★★★
- ✅test ★★★★★
- 🗂Doubts about the topic 2 ★★★★★
- ✅Test 3 ★★★★★

▼ **Block settings**

Allow geolocation of the ☑
users.

▼ **Where this block appears**

Original block location   Course: Communications Theory
⑦

Display on page types   [Any page ▼]
Default region ⑦   [Left ▼]
Default weight ⑦   [-10 (first) ▼]

▼ **On this page**

Visible   [Yes ▼]
Region   [Left ▼]
Weight   [-10 (first) ▼]

[Save changes]   Cancel

Figure 4.5   Configuration options of the INTUITEL block.

For the options "Display on page types", the option "Any page" must be selected so that the block appears in any page of the course. It is recommended to make the INTUITEL block very visible by situating it in high priority positions so that TUG messages and LORE recommendations are easily noticed by the student. Last, the teacher can select if geolocation of the student will be registered or not.

4.2.2 IMC Learning Suite

Uta Schwertel and Sven Steudter

The INTUITEL services have been integrated into the Learning Management System IMC Learning Suite (ILS)[2]. The IMC Learning Suite is a comprehensive commercial e-learning solution for the planning, implementation and

---

[2]see https://www.im-c.de/en/learning-technologies/learning-management/ (Last accessed on 15 August 2016).

management of company-wide professional learning and development. The learner interacts with the INTUITEL enhancements (recommendations and tutorial guidance) through an IMC Learning Portal. The Learning Portal offers a lightweight, responsive and easy-to-use front-end that allows learners personalized access to online courses on different end-user devices.

Through the Learning Portal a learner can find and book an online course, and then learn with the course materials (e.g. watch videos, work through Web-based-trainings, read PDFs or other documents, perform multiple-choice tests, discuss with peers about the course content, order certificates etc.). Tutors and administrators use the more powerful IMC Learning Suite administrator view to create courses, supervise courses, administer participants or manage certificates.

For the INTUITEL enhancements we utilized and extended the Universal API of the Learning Suite (4.6). The INTUITEL specific settings (e.g. Service Point IP for using INTUITEL services) and other technical configurations are defined by an administrator in two XML based configuration files. This makes INTUITEL Learning Suite installations easily reproducible on other servers.

An INTUITEL enabled scenario for the IMC Learning Suite requires the following components: an INTUITEL enhanced Learning Portal connected to a Learning Suite, an INTUITEL back-end installation connected to the Learning Suite and Learning Portal, and an integrated online course annotated with INTUITEL specific pedagogical metadata.

In the previous Moodle example we explained in detail the different messsage endpoints (Learner Update, USE, TUG, LORE, etc.). In contrast to the following paragraphs focus on the access to these components from a learner and a tutor/teacher perspective. The overall work flow behind this is similar for all extensions. Depending on the LMS functionality there may be slight differences.

4.2.2.1 Course Creation in IMC learning suite

Defining and creating an INTUITEL enhanced course in the IMC Learning Suite involves the following three main steps:

1. Perform Instructional Design and Content Creation: An expert defines the instructional design of the course and provides the content elements (Knowledge Objects) of the course (texts, videos, audios, WBTs, tests, etc.). The example course "Advanced Java Concepts" for the INTUITEL enhanced Learning Portal has been created by the INTUITEL project partner University of Reading. A graphical overview of the instructional design of the example course can be found in Section 4.4.1 of this book.

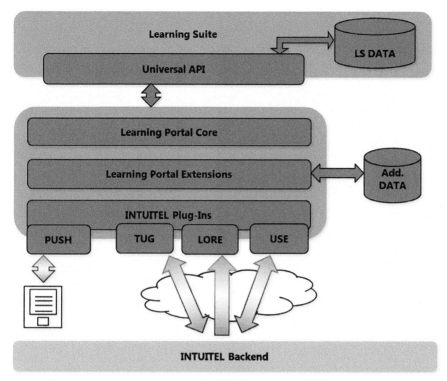

Figure 4.6 Architecture of the INTUITEL enhanced IMC Learning Suite.

2. Create Course in IMC Learning Suite: A course administrator creates the course in the INTUITEL enhanced Learning Suite and uploads the learning objects (Knowledge Objects) for the course. The Learning Suite can be configured as to which meta-data are given within the Learning Suite and which metadata have to be added through the external INTUITEL editor to each learning object. We have chosen to keep the meta-data that have to be added within the Learning Suite sparse and conform to standard non-INTUITEL enhanced courses. Currently available meta-data for each Knowledge Object within the Learning Suite are:

- Name
- Description
- Duration
- Keywords
- Can be used in external applications (yes/no)

- File upload
- Preview image
- Player settings (parameters to configure the internal player)

Note that this set of data is specific for the IMC Learning Suite. Other LMS plugins may provide different data.

3. Annotate Course with INTUITEL Editor: In order to be able of recommending learning steps, INTUITEL needs to have data about the pedagogical design of the course. This data is given outside the Learning Portal and Learning Suite by means of the INTUITEL Editor. This piece of software is an installable program that allows the tutors or teachers to draw Learning Paths by connecting and grouping the Knowledge Objects of the course. The INTUITEL editor is available from the INTUITEL website[3].

To make these annotations, the expert configures the INTUITEL Editor for the respective Learning Management System (with specific configuration data that are given to the expert) and loads the knowledge objects created in the Learning Suite into the Editor. The expert can now define the instructional design in the editor and annotate the objects through the editor.

4.2.2.2 Accessing courses through IMC learning portal

After having finished the above described steps, the course is finally ready for delivery through the IMC Learning Portal. The following paragraphs describe how a learner accesses the INTUITEL enhanced course.

1. Access to INTUITEL enhanced Learning Portal: The learner accesses the landing page of an INTUITEL enhanced Learning Portal. By clicking on "Courses" learners get a list of all available courses in the course catalogue of the system.

2. Sign-up to INTUITEL Learning Portal: The learner has to sign-up to the system to be able to book and use an INTUITEL enhanced course. At this stage it is also possible for the learner to edit his or her user profile data.

3. Book and start the INTUITEL enhanced course: After registering for the course a learner can start the course. The learner will then see the course elements with the main concept containers (CC) and knowledge objects of the course. This means a user can also browse in the course without using the INTUITEL recommendations. Figure 4.7 shows the start page of a course

---

[3] see http://www.intuitel.de/public-software/ (Last access: 17 November 2016).

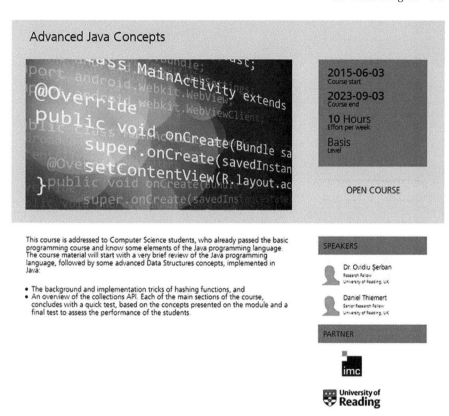

Figure 4.7 Starting Page of a course in IMC Learning Suite and possibility to open course.

whereas Figure 4.8 shows the course curriculum overview as the learner sees it after he or she has booked into the course.

4. Start the course: A learner starts working on the course by clicking for example on the Knowledge Object "Advanced Java Concepts". This will open a detail page where on the right side the learner can work through the material, and on the left side the learner sees boxes where INTUITEL tutorial guidance (Figure 4.9) messages or learning object recommendations (Figure 4.10) will show up.

5. Receiving Learning Recommendations in IMC Learning Portal: As explained in Sections 3.4 and 3.5, learning recommendations are sent to the LMS plugin in the form of TUG and LORE messages. The learner can interact with the messages in the Intelligent Tutor by answering questions or following recommendations shown in the boxes. Depending on the selections, different

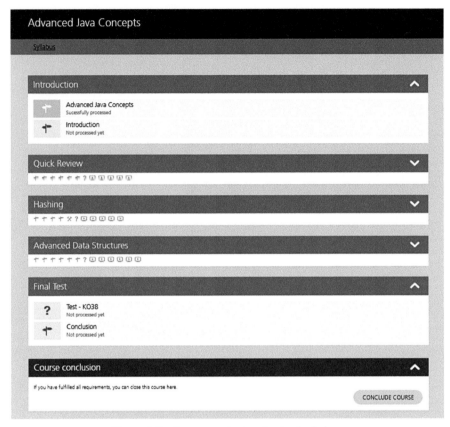

Figure 4.8   Course curriculum in standard view.

recommendations are calculated by the INTUITEL back end. An example run through the INTUITEL example course is described in Section 4.4.1.

Tutorial Guidance (TUG): Through a tutorial guidance message a user can select which Learning Pathway he or she prefers. Other types of tutorial guidance messages ask for other preferences or personal data, e.g. age or preferred language (see Figure 4.9), or simply state that currently no guidance is available or simply state that currently no guidance is available.

Learning Object Recommendations (LORE): In a Learning Object Recommendation message (LORE) the system recommends next relevant learning objects when a user is at a certain state in the learning path. If the system also sends out weightings of the recommendations this is displayed by coloring

Figure 4.9 Example Tutorial Guidance (TUG) – personal preferences.

the scale chart below the recommendation as illustrated by Figure 4.10 (the width of the chart correlates to the weight of the recommendation).

4.2.2.3 Summary

We have shown how the INTUITEL services have been integrated into the commercial Learning Management System IMC Learning Suite and how the recommendations and tutorial guidance are delivered through the IMC Learning Portal. We have conducted experiments to check if the INTUITEL tutorial guidance improves the user experience and quality of learning. For the IMC Learning Suite these tests have been conducted under the lead of the University of Reading. Users stated e.g. that the INTUITEL enhanced Learning Suite offers great potential for improving Open Online Courses towards a more personalized and motivating user experience. To demonstrate the power of the INTUITEL recommendations more courses with a complex pedagogical modeling will have to be developed and presented to users.

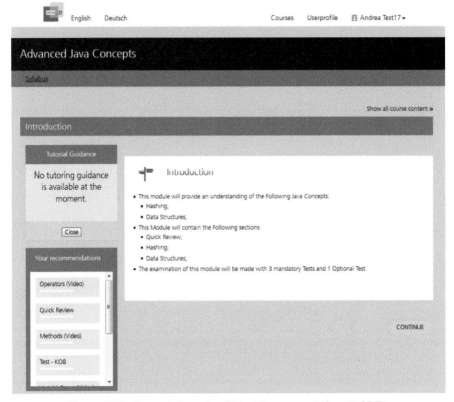

Figure 4.10   Example Learning Object Recommendations (LORE).

4.2.3 Ilias

Kevin Fuchs and Peter A. Henning

ILIAS is a free open source Learning Management System. It provides an integrated environment for sharing, structuring and organizing learning content. Moreover, it provides functionality for communication and collaboration among learners and teachers. It is also capable of tracking a user's progress.

To integrate the ILIAS LMS into the INTUITEL system, the LMS had to be extended by a plugin that comprises two components: a REST interface to communicate with INTUITEL and – behind the REST interface – a compontent that retrieves the desired data related to users and learning content from the LMS. The following will briefly describe the implementation of the plugin and how it communicates with the INTUITEL system.

4.2.3.1 Plugin implementation

The plugin is based on a ILIAS-skin that can be activated or deactivated by the user. In his personal settings the user may select an INTUITEL skin instead of the ILIAS default skin. So the user may decide himself if he wants INTUITEL functionality while working with the LMS. In contrast to the IMC Learning Suite, no separate activation of the INTUITEL functionality for a single course is needed. The ILIAS plugin sends Learner Updates for any learning object the learner accesses without caring if the respective course has been annotated. In the case of unannotated learning content, the INTUITEL back end will simply perform a recommendation process with no result. On the one hand this enforces a recommendation process even if it is not needed. On the other hand it makes the plugin more lightweight.

The communication between the plugin and the INTUITEL communication layer runs asynchronously in the background. Hidden from the user, Learner Update messages as described in Section 3.5 are sent to the INTUITEL communication layer. This happens each time a learner enters a learning object or finishes a test.

The Learner Update initiates the respective message sequences (see Section 3.4) to trigger the recommendation process in the INTUITEL back end. The result of this process is sent from INTUITEL to the REST interface of the ILIAS plugin in the form of TUG and LORE messages. The plugin then injects this information into the ILIAS web frontend, displaying a info box to the user. These messages may contain Learning Object recommendations (LORE) in the form of ranked lists of other related learning content or Tutorial Guidance messages (TUG) which may be textual learning recommendations, and interactive forms with which the user is asked some questions in order to refine the recommendation process or let the user choose from multiple recommended learning pathways. Figure 4.11 shows how a LORE message looks like in the web frontend of ILIAS.

In general, ILIAS offers two ways to retrieve data from its database. First there is a SOAP Web Service interface by which data can be read and manipulated. This is most interesting for external applications that want to access the ILIAS database from outside. Second, ILIAS offers a plugin architecture with which new functionality can be easily installed from the ILIAS administration back end. In the case of the INTUITEL system we decided to use the internal plugin mechanism of ILIAS since only this way we could inject the INTUITEL messages into the ILIAS frontend seamlessly. Furthermore, this made it easy to implement the beforehand mentioned skin functionality.

Figure 4.11 A LORE message in the ILIAS web frontend (window on the right).

The plugin is managed from the administrations back end and needs only two parameters: the URL to the REST interface of the INTUITEL communication layer and a local port on which the REST interface of the plugin is supposed to run and which is addressed by the communication layer. The plugin, which is implemented in PHP, installs an integration component and the ILIAS skin. The integration component that is written in Java contains a minimally configured standalone tomcat server that is started by the plugin automatically after installation. The tomcat server listens on the port that is specified by the installation parameter "server port" and runs the REST interface for the communication between the plugin and the INTUITEL system. The REST interface consists of the endpoints explained in Section 3.5.

The skin consists of two variants the user can choose from – one for a popup box and one for a box that is embedded into the ILIAS GUI. Both the popup box and the embedded one are created by a JavaScript that accesses the DOM of the HTML structure. The script simply injects the HTML code of the INTUITEL box into the HTML of the ILIAS GUI.

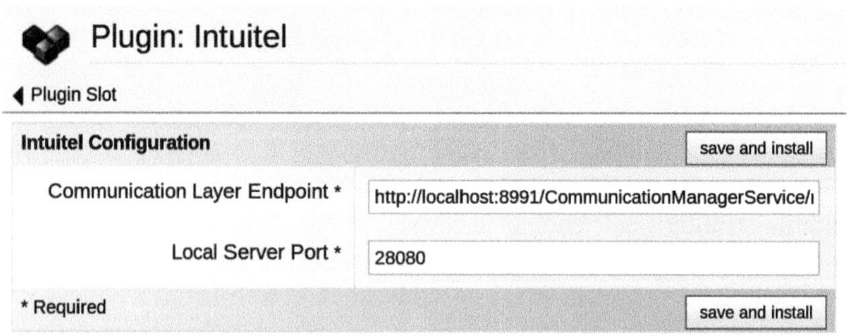

Figure 4.12 Configuration of the plugin in the ILIAS administration back end.

4.2.3.2 Learner tracking

In the first instance the tracking of a learner's behavior is performed by a JavaScript function. That function takes the URL of the currently opened ILIAS page that contains the id of the respective learning object presented on that page. By using asynchronous JavaScript calls (Ajax) this id is sent to the REST interface of the integration component waiting asynchronously for a response. Figure 4.13 shows the data flow between browser, integration component and the INTUITEL system.

While the JavaScript function only tracks the object ids the integration component contains the entire logic to communicate with INTUITEL, as well as to collect and process data from the ILIAS database. As a first reaction of being informed about a learner's action the integration component sends a Learner Update message to the communication manager of the INTUITEL back end. INTUITEL then in turn sends a USE request (see Section 3.5) to the integration component to collect user-specific environment and performance data. To generate this data the integration component accesses the ILIAS database.

In the last step, when the integration component has responded to the USE request, the INTUITEL back end starts its recommendation process and sends the result to the integration component in the form of TUG/LORE messages. Having received that TUG/LORE message the integration component satisfies the original REST request of the JavaScript function by responding with the TUG/LORE content. The JavaScript then displays the TUG/LORE content in the INTUITEL info box within the learner's browser. The way the plugin communicates with the INTUITEL system is due to the push scenario as described in Section 3.5. This means, that once a Learner Update has been

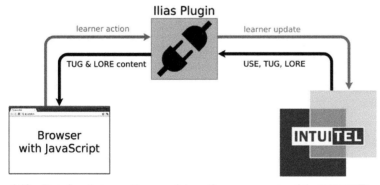

Figure 4.13 Data flow between Browser, integration component and the INTUITEL system.

sent actively by the LMS, then the LMS plugin remains completely passive, only reacting on USETUG, and LORE messages of INTUITEL.

4.2.3.3 Architecture

The inner architecture of the integration component consists of three main components that are illustrated by figure reffig-ILIAS-plugin-architecture and explained in the following.

Communicator   The Communicator builds the REST interface with which the INTUITEL communication layer and the JavaScript functions in the user's browser communicate. The Communicator is exchangeable and could also provide any other communication method than REST. being absolutely unaware of the content, sources and destinations of the messages it processes, the only task of the Communicator is the provision of the REST interface as well as passing and receiving messages from and to the layer on the next level which is the Broker described below.

Broker   The Broker represents the second layer of the design. Its task on the inbound side is to receive messages from the Communicator and delegate them to the appropriate Core module which can be the TUG, LORE, USE and the Learner Update module, which are described below. This means that–in contrast to the Communicator–the Broker knows about the destinations and sources of messages in order to determine the fitting Core module to which it delegates the message. However, the Broker is indifferent to the contents of messages. On the outbound side the Broker takes messages from the Core components and delegates them to the Communicator.

Core   The Core contains the modules that are in charge of handling TUG, LORE, USE messages and Learner Updates. On the outbound side, each of them receives corresponding messages from the Broker. They read the message payloads and take appropriate action. On the outbound side they delegate TUG, LORE, USE and Learner Update messages to the Broker which in turn passes them to the Communicator. The Communicator finally passes the messages either to the INTUITEL Communication Manager or to the user's browser. The Core also contains modules that provide general services for the TUG, LORE, USE and Learner Update modules. Such general services are:

- Marshaling services to transform XML payloads for processing
- Database connectivity to retrieve data from the ILIAS database
- Error handling
- Logging functions

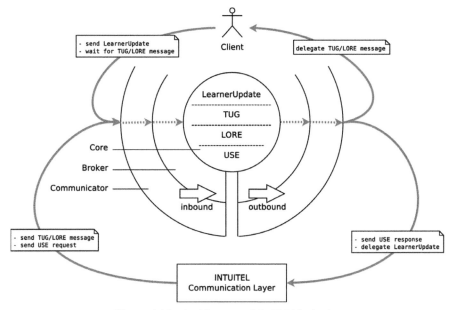

Figure 4.14 Architecture of the ILIAS-plugin.

4.2.3.4 Maintainability

Encapsulating the above described methods in separate services contributes to the ability to replace and modify the according functions without affecting other modules. One could – for example – use other data formatting than XML or migrate to another database. A quite realistic scenario is the occurrence of changes of the ILIAS database scheme, i.e. due to migrations to newer versions. Hence, placing such functionality into separate components contributes to the maintainability and extensibility of the entire system.

4.2.4 eXact learning LCMS

Elisabetta Parodi

eXact learning LCMS[4] is a commercial Learning Content Management System, supporting instantaneous, company-wide collaboration for the creation of standard-compliant, reusable and easily maintainable learning content. Furthermore, the LCMS allows single-source multiple outputs

---

[4]http://www.exact-learning.com/exact-learning-lcms/

publishing scenarios. The LMS part of it, allowing content delivery, supports course offer and tracking of learners' progress.

In the context of the INTUITEL project, eXact LCMS was extended to interact with the INTUITEL back end in order to send users' data and information about progress in courses and therefore be able to receive INTUITEL related tutoring and guidance messages. Of course, to make a course to be assisted by the INTUITEL tutor, the course needs to be previously annotated, i.e., by the INTUITEL Editor. In the following we present shortly the eXact learning LCMS, how it was extended to exchange communications with the INTUITEL back end and the user experience with the INTUITEL-enhanced interface of the LMS.

4.2.4.1 About eXact learning LCMS

eXact learning LCMS is an industry reference Learning Content Management System that responds to today's varying business pressures, supporting instantaneous, company-wide collaboration for the creation of critical learning content. Developed by eXact learning solutions[5], an Italian SME, eXact learning LCMS maximizes a company's existing content investments, while supporting learning content strategies that improve key business processes. Figure 4.15 shows an overview of the eXact learning LCMS.

Based on XML technology, eXact learning LCMS is designed for the extensive reusability of content chunks. Moreover, it supports SCORM 1.2 & 2004, IMS and Tin Can xAPI standards, allowing single-source multiple output publishing scenarios. eXact learning LCMS allows for rapid content

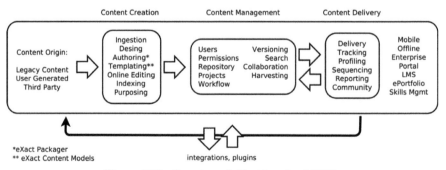

Figure 4.15   Overview of eXact learning LCMS.

---

[5]http://www.exact-learning.com/company/

development through the seamless collaboration of subject matter experts and course designers, regardless of their location and specialization. This provides cost-effective, repeatable and consistent content production projects.

The LMS part of the platform allows courses management and delivery. Courses are offered via a catalogue. There are several modalities supported for enrollment, from self-enrollment to teacher's enrollment. Once registered for a course, a learner can access its contents, with an interface similar to the one reported in Figure 4.16.

As from Figure 4.16, information about the whole course is presented on top of the page, while course materials are listed below. For each material, information about its completion is provided by the system (i.e. Incomplete, to be started, etc.). Other information about course staff and enrollment is presented on the right side. The LMS is able to track the learner's interaction with standard-compliant courses. Such information is forwarded to the INTUITEL back end via the developed extension and, together with other information, supports the reasoning for recommendation and guidance.

4.2.4.2 Extension

eXact learning LCMS was put in communication with the INTUITEL back end by means of additional, pluggable code, that we call here "extension". This extension provides on one side back end communication with the INTUITEL services, and on the other side windows added to courses on the front end, supporting the dialogue between the learner and the INTUITEL reasoning. The set-up of the extension is very easy, just creating an IIS virtual directory to the folder containing the code of the web services.

Figure 4.16 Sample course interface.

The back end part of the extension is implemented as a set of REST C# web services using Web API Framework (NET Framework 4.5). The choice of such technology is because the eXact LMS is developed in Microsoft C#. The communication is HTTP based and the server side implements a requestresponse message system, based on XML format, through the implementation of many specific end points each one of which manages one type of message.

The REST end points and the messages accord to the definitions of Section 3.5. The Web Services are developed in C# as a WEB API project; this can give more flexibility and more compatibility as a standard .Net Web Service. The web services wait for a message from the INTUITEL system and then send an answer back. The Object model has been created using Xsd2Code9 starting from the supplied XSD. Each object (e.g.: Learners, LMSProfile, etc.) implements two specific methods (e.g.: Serialize, Deserialize) used to marshal and transmit the object representation between the communication endpoints. Also the extraction of the data from the eXact LCMS that are requested from INTUITEL is implemented directly in these objects and not in the Web Services, so it possible to use the developed object library also with client applications different from INTUITEL and other communication layers.

For sending Learner Update information we configured the INTUITEL back end for the special communication style "Push with Learner Polling" (see Section 3.5): We implemented the "learners" method assuming that INTUITEL will periodically make calls to the LMS and the LMS returns a list of active users and their activity, meant as courses accessed by each user since previous "learners" call. For the front end, some javascript code manages the windows dealing with the learner-INTUITEL dialogue. The design choice for the user interface was a compromise between the need to allocate part of the screen area for dedicated messages, and the wish to keep the interface neat and essential. Therefore, the adopted solution was to have expandable pop-up windows that can be minimized once the related message has been read.

4.2.4.3 Demonstrator

As previously indicated, our demonstrator contained a course, linked with pedagogical metadata (before being actually delivered to learners) thanks to previous annotation via the INTUITEL Editor. From the learner's point of view, three small frames are added to the default interface, as shown by Figure 4.17.

Entering this course, the first evident novelty respect to non INTUITE-Lenhanced courses is the appearance of three small labels on the left side,

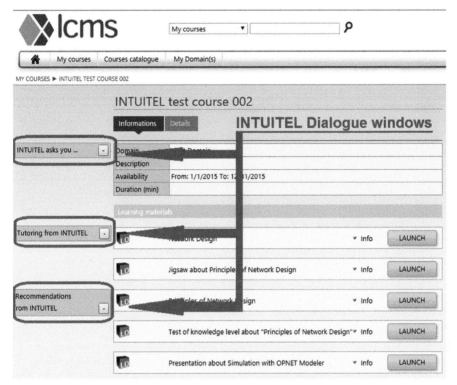

Figure 4.17 Sample INTUITEL-assisted course in eXact.

as highlighted in Figure 4.17. These labels can be expanded into small windows, which manage the communication between INTUITEL and the learner. Learners can access and play courses as usual. What is different with the INTUITEL extension is the appearance of the three labels on the left side. These labels can be expanded into small windows managing the communication between the learner and the INTUITEL core. The (expanded) appearance of these windows is presented in Figure 4.18. As illustrated there, INTUITEL-enabled courses in eXact present the following three collapsible windows:

1. A first window presents a few questions to the learner to store information that is not directly retrievable from the basic profile stored by the LMS, such as the daily mood, preferred interaction kind, etc.
2. A second window takes care of the tutoring dialogue, this corresponds to TUG messages. The TUG interface carries out a dialog between the

learner and the INTUITEL system. A TUG Interface must be present in any INTUITEL enhanced system (see Section 3.5 for further description of the TUG interface).

3. The third window provides INTUITEL recommendations about the suggested learning path for the current learner; this corresponds to the LORE interface. The LORE interface makes a recommendation by the INTUITEL system known to the learner. It is, in its simplest implementation, a special case of the TUG interface and therefore present in any INTUITEL enhanced system.

4.3 Compatibility to Existing Learning Formats

Luis de-la-Fuente-Valentin and Daniel Burgos

There are many competing standards for content production and package in the current learning market. For example, SCORM importing and exporting is supported by many Learning Management Systems. Another example is IMS Learning Design, also based on the IMS Content Packaging specification but with a more expressive vocabulary to define learning pathways.

With this technological landscape, with a lot of content already produced in different formats, it is realistic to think of end users as reluctant to shift from their preferred format to SLOM. Furthermore, INTUITEL enabled content requires an edition phase (the enhancement of content with semantic metadata) that may hinder the adoption of the framework.

In order to reduce the INTUITEL adoption gap and make it easy to teachers to migrate from their preferred format to INTUITEL, the framework offers editing capabilities also enhanced with a translation system called the INTUITEL Merger. The Merger is able to translate learning materials from

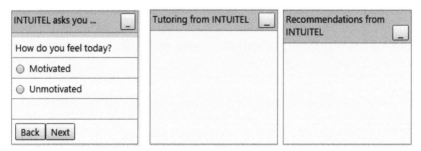

Figure 4.18   Windows for INTUITEL dialogue in eXact LMS.

and to different learning formats, including SLOM. This section offers a technological perspective of the Merger approach and discusses the lessons learned on the implementation of the translation from and to the selected formats, namely IMS LD, Semantic Media Wiki, SCORM and SLOM.

4.3.1 The Transformation Approach

The mantra *write once, run anywhere* was initially proposed by Sun Microsystems as a marketing slogan applied to software source code, specifically the Java virtual machine infrastructure that pursues cross-compatibility among operating systems. This idea of reusing writing efforts conceived for software production has also been adopted in the eLearning field. In other words, as courseware gains more features, it also gains more complexity and, consequently, it is harder to achieve cross-platform compatibility.

Courseware production is a time-consuming, difficult task that requires a great deal of effort from practitioners. Therefore, practitioners may expect their content to be reusable (i.e., may be used more than once) and interoperable (i.e., possible to use on different platforms), but this is not always the case.

This is why interoperability and reusability are always present in any eLearning development. Many standardization efforts have been implemented with different levels of success. A good example is the SCORM framework, whose packaging format has reached the status of a de facto standard in courseware packaging. Many platforms are able to import SCORM packages, while few of them are able to exploit the advanced SCORM features for dynamic content.

New developments must decide whether to be compliant with older formats or to build a completely new paradigm. In the first case, they will inherit the limitations of the older formats. In the second case, new systems will be unable to use a large amount of course content already created for older systems.

Since the SLOM format is an example of this second case, it would be desirable to find a method to SLOM-enhance old course content. The approach taken in the INTUITEL project was to develop a transformation methodology, supported by the INTUITEL Merger as a software transformation tool. In other words, the SLOM format does not inherit restrictions from older content production systems, and provides methods to enable the reuse of older course content.

Many previous works use transformations between formats to achieve course content reusability and interoperability, or even to facilitate new content production paradigms. On the most basic level, the way to translate unstructured learning content into structured learning objects is described [22]. The authors analyze the process of learning object creation and the transformation methods, identifying possible pedagogical issues.

The intrinsic limits of a given framework may also lead to the inclusion of a transformation process in the course life cycle. For example, the inherent complexity of the IMS Learning Design specification hinders the course authoring process, and IMS-LD editors have not achieved the expected ease of use. Certain researchers [6, 51] propose to start the authoring process with well-known techniques used in software production and use transformation tools to generate IMS-LD compliant courses. Barchino et al. [97] follow a similar approach by applying software creation techniques to create SCORM course content. A posteriori analysis of the results reveals an interesting fact: a successful tool and a quality transformation are not enough, since course authoring has pedagogy-specific issues that cannot be addressed by non-pedagogically designed tools. Therefore, it is important that humans complete the process in order to apply the pedagogical touch.

The SCORM format has achieved great success, while it also has some deficiencies. For example, the extensive cataloguing by means of metadata is an obstacle for developers and vendors because the existing content has to be provided with metadata to be SCORM compliant [96]. To increase the interoperability of digital libraries with existing platforms, Arapi et al. [2] developed an architecture that relies on format transformations to interconnect systems.

A review of the literature highlights the relevance of the interoperability and reusability of learning content, and justifies the use of transformation tools to achieve these goals. The following lessons learned have guided the design and implementation of the SLOM transformation tool:

- Transformations may introduce pedagogical problems, as identified in earlier works.
- Well-known tools can be used to improve the authoring experience. Although results are good, pedagogical issues are still present.
- A well-designed transformation tool and methodology enables reusability of successful content in new frameworks.

Different formats for eLearning content production and packaging pursue different goals and have been constructed based on different learning theories. For example, IMS Learning Design claims to support collaborative learning and therefore includes vocabulary to model roles, support activities, forums, groups, etc. However, this is not the case with SLOM, which interprets learning material as something to be individually consumed and takes into account peer activity in order to elaborate recommendations. The same applies to other formats such as SCORM, LOM, IMSCP and AICC. These have been designed on a different basis, and it is therefore not possible to produce a lossless translation between two of these formats.

To overcome this problem, the decision of the INTUITEL team was to support a semiautomatic transformation process, where most of the tasks are automatically carried out by the software, while fine-grain details require human intervention in order to be completed. The complete flow is as follows:

1. The course author identifies a course whose content is adequate for a given knowledge domain, but not written in SLOM format.
2. The course author uses the INTUITEL Merger to generate a SLOM version of the course. This produces a SLOM package.
3. The course author then imports the SLOM package into the INTUITEL Editor and manually completes the transformation.

After this process, the course content is an INTUITEL-enabled course to be used within any INTUITEL-enabled platform. Within this approach, the researchers recognize the impossibility of designing a completely automatic transformation process, while acknowledging the relevance of the details of the course content.

Furthermore, the implemented transformation is bidirectional. That is, SLOM can also be translated into other formats. Such SLOM-to-other-formats transformation follows the same semiautomatic approach, but the other-format-editor required to complete the third step is not provided as part of the INTUITEL framework.

To maximize interoperability and reusability, the INTUITEL team has chosen the best-known eLearning formats on the current market: SCORM and IMS Learning Design. Furthermore, Semantic MediaWiki courses can also be imported and translated into SLOM. Since Semantic MediaWiki (SMW) is not built on a predefined schema nor imposes assumptions regarding the schema used for the description and annotation of learning content, the SLOMto-SMW transformation solution is based on explicit import and mapping declarations that determine how PO and SLOM elements are mapped to Semantic MediaWiki-specific elements and vice versa. These mappings have to be declared on a case-by-case basis since Semantic MediaWiki is per design an ontology-based authoring tool for arbitrary data and domains rather than a prescriptive schema for describing eLearning content such as IMSLD or SCORM. This aspect is fundamental to the SLOM-to-SMW transformation specification. The architectural software approach allows for the inclusion of new formats for translation. The next section briefly describes the implemented transformations.

4.3.2 Implemented Transformations

4.3.2.1 SCORM

The SCORM format stands on the basis of the IMS Content Packaging specification, and therefore uses organization, resource and item as organizational units. The mapping rules take the top-down approach and therefore start by reading the organization elements and hierarchically read inner elements until resources are read.

In the transformation process, each organization found at the SCORM manifest file results in the creation of a CC in the corresponding SLOM file. Each property available in SCORM is transferred as an individual annotation to the CC. Then, items (and related resources) available at the organization are translated to Knowledge Objects in the SLOM file.

Last, sequencing rules are created in the SLOM file. The purpose of INTUITEL regarding a priori sequencing rules is much simpler than in SCORM, because INTUITEL sequencing is calculated at runtime. Thus, the resulting SLOM file has a single Macro Learning Pathway, and each Concept Container has a micro Learning Pathway that contains all the inner Knowledge Objects.

4.3.2.2 IMS Learning Design

The IMS Learning Design also stands on the basis of IMS Content Packaging, and establishes complex sequencing rules that build the structure of the course. The translations read these sequencing rules and map them to the different elements available in the SLOM format.

The procedure follows a bottom-up approach: in the first step, actual learning content is identified (i.e. learning objects inside environments, related to learning activities) and a collection of Knowledge Objects is created. Then, these KOs are wrapped into CC according to the sequencing rules of the IMS-LD course. That is, all the KOs identified in the same act are related to the same CC. Next, the CCs are sequenced into Macro Learning Pathways (MLPs). Finally, the micro Learning Pathways (mLPs) inside the CCs are created. In a typical case, an imported course will have a single MLP and multiple mLPs, one for each CC.

4.3.2.3 Semantic MediaWiki

Semantic MediaWiki (SMW) does not define any new canonical data or description format since the logical model that builds the basis of its knowledge representation formalism is to a large extent based on the Web Ontology Language (OWL). The transformation process is based on the definition of mapping declarations that need to be built for the purpose of a given course, that associate course properties (i.e. Semantic Media Wiki properties) with CCs and KOs.

4.3.2.4 Integration in a common format: INTUITEL merger

The implemented transformations have a very different nature. For example, both SCORM and IMS-LD are based on the IMS Content Packaging specification but, due to their intrinsic characteristics, SCORM transformation is based on XSLT while IMS-LD is built on top of a Java-based procedure. Furthermore, SMW does not offer the concept of a concept package, and the SMW courses are stored in a web server.

Due to all these differences in structure and underlying technologies, a common layer is provided to unify the transformation interface and ease the process. Such a layer, the INTUITEL Merger, is a piece of software able to import from and export to different eLearning formats. The transformation process follows a semiautomatic approach, where fine-grain details require human intervention to be completed.

4.3.2.5 Architectural principles

The purpose of the INTUITEL Merger is twofold. First, its functionality enables the inclusion of already existing course material in the authoring process of INTUITEL-enhanced courses. Second, it enables the exportation of course content in different formats.

As shown in Figure 4.19, the SLOM format plays a central role in all the transformations, that is, whenever an import is executed the result of the importation is an internal representation of the course as SLOM format. This design decision emphasizes the relevance of SLOM as a central part of the INTUITEL Framework. In addition, the structure of the Merger enables non-SLOM transformations. That is, the user can use the Merger to obtain an IMS-LD representation of a Semantic MediaWiki course by just importing and then exporting the course using the proper parameters. With this SLOM-centric procedure, the Merger includes SLOM-to-SLOM importation and exportation, thus enabling the transformation from SLOM files to other formats. According to Figure 4.19, in the first released version of the tool, four formats are

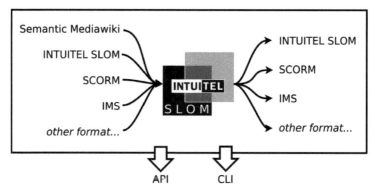

Figure 4.19   Transformations in the INTUITEL merger.

supported (SCORM, IMS-LD, SMW and SLOM). As the implementation of the transformations is plugin-based, the tool is extendable with new supported formats.

4.3.2.6 Merger use cases: Standalone or integrated

The INTUITEL Merger provides an interface for two main use cases: its use as standalone tool and its integration on larger software. In the INTUITEL case, the Merger has been integrated as part of the Editor. Therefore, the user can complete the authoring life cycle. That is, to import a course from an external file, load its SLOM representation into the Editor, manually edit finegrain details or even modify the structure and export to the desired format.

A built-in command line interface enables the use of the INTUITEL Merger as a standalone tool. In this case, the transformation is an end-to-end process that receives an input file (or a URL for SMW imports) and produces an output file in the specified format. Here, the internal SLOM representation is a temporal placeholder that bridges the transformation process, while the end user has no access to such internal representation.

4.4 Sample Courses

To demonstrate the functionality of the INTUITEL prototype, we present three example courses in this section. Note that the creation and structuring of a course is entirely independent from the concrete LMS keeping the learning material. The main task of a course creator is the semantic annotation of the course material on the meta-level according to the Pedagogical Ontology as described in Section 3.1. The concrete LMS content is then linked to the resulting SLOM data (see Section 3.6).

The variety of topics and LMS that we chose for the sample courses especially emphasizes the Plug-and-Play principle by which meta-knowledge can be applied to any learning material of any knowledge domain. Therefore, the following sample courses were all created following the same procedure and we will set a different focus on each of them.

We first start with the course "Advanced Java Concepts" tested with the IMC learning suite. With this course we will focus on the guidance as it is experienced by the user on the screen. The second course – tested with Moodle – is about network design and we will focus on the subdivision of the learning material into Concept Containers and Learning Pathways. The third course was tested with ILIAS and provides an introduction into Albert Einstein's theory of Special Relativity. With this course we will have a closer look at the design of micro and Macro Learning Pathways.

4.4.1 Advanced Java Concepts

Uta Schwertel and Sven Steudter

To test the INTUITEL recommendations with the IMC Learning Suite the INTUITEL partner University of Reading has designed and piloted an online course "Advanced Java Concepts". The course addresses Computer Science students who already passed a basic programming course and know some elements of the Java programming language.

4.4.1.1 Course design

The course starts with a brief review of Java programming language, followed by some advanced Data Structure concepts, implemented in Java: the background and implementation of tricks and hashing functions and an overview of the collections API. Each of the main sections of the course concludes with a quick test, based on the concepts presented in the module and a final test to assess the performance of the students.

The course offers knowledge objects using different media types (video, text, test). For example, the concept "Operators" is explained in text form or alternatively as a video. Furthermore, the course is level oriented and distinguishes between introductory and advanced material and the learner can add preferences regarding the level of detail or the preferred media type. In Figure 4.20 a graphical description of the pedagogical design is given. The elements of the course are summarized in Table 4.2.

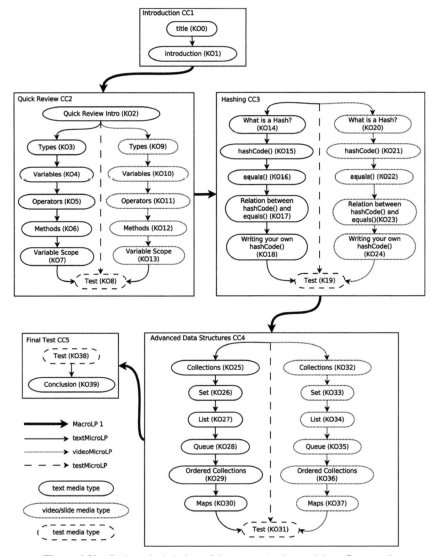

Figure 4.20   Pedagogical design of the course "Advanced Java Concepts".

4.4.1.2 Example run through the course

In the following section we describe a brief example run through the course "Advanced Java Concepts". We assume a newly registered user has opened the "Advanced Java Concepts" Knowledge Object in the course. Through a Tutorial Guidance message triggered by the INTUITEL system the user is then asked for the preferred Macro Learning Pathway, namely, whether he or

Table 4.2 Example elements of course "Advanced Java Concepts"

INTUITEL Structural Concept	Example Elements in Course
Knowledge Domain (KD)	Advanced Java Concepts
Concept Containers (CC)	Introduction (CC1)
	Quick Review (CC2)
	Hashing (CC3)
	Advanced Data Structures (CC4)
	Final Test (CC5)
Knowledge Object (KO) Types	Text, Video, Test
Knowledge Objects (KO)	[KO0] to [KO39] (see Figure 4.20)
MacroLPs	MacroLP1: Full course (all materials)
	MacroLP2: Advanced concepts only
MicroLPs	textMicroLP: I prefer text materials
	videoMicroLP: I prefer videos
	testMicroLP: I prefer tests

she wants to do the full course with all materials or the advanced concepts only. The learner selects in the drop-down list the option "Full course (all materials)", viz. chooses the MacroLP1 (Figure 4.21). In a next step the user

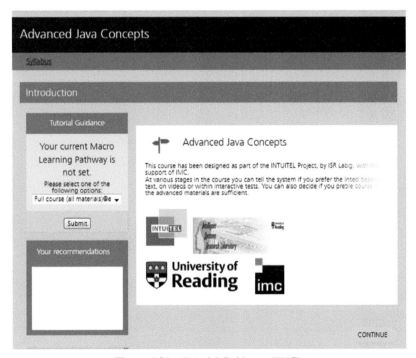

Figure 4.21 Tutorial Guidance (TUG).

is asked for the preferred MicroLP, viz. whether he or she prefers texts, videos or tests only.

Suppose, the user chooses "I prefer videos". The modeling of the course does not yet offer alternative Knowledge Objects at this stage which is why the system first recommends the Knowledge Object "Introduction" to the user. This Learning Object Recommendation (LORE) appears in a box where the user can select the recommended learning object (Figure 4.22).

Suppose the user follows the recommendation. This opens the Knowledge Object "Introduction", which the user can consume, and triggers further Learning Object Recommendations that take into account previous selections of the user regarding the preferred Macro- and Micro Learning Paths, viz. the user is presented preferably videos and is recommended also Learning Objects of Concept Container 2 (CC2: Quick review).

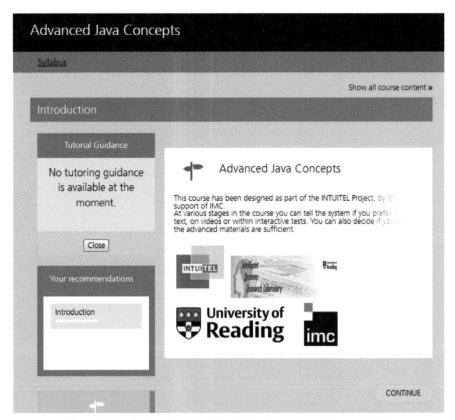

Figure 4.22   Follow first Learning Object Recommendation (LORE).

In our example the user selects the Knowledge Object "Operators (Video)" and can view the video in the right window of the screen. The INTUITEL system calculates further recommendations and displays them in the left box "Your recommendations". Again, the user is free to choose what to select next, but ideally follows the recommendations provided through the INTUITEL system. The user chooses to do the test (KO8) and is presented the interactive test provided by the IMC Learning Portal (Figure 4.23). The test is evaluated by the Learning Management System and the result (passed/failed) is presented to the user. In the current setting the user can, however, redo the test.

After completing the test the user selects further recommended Learning Objects which now belong to the next Concept Container "Hashing (CC3)".

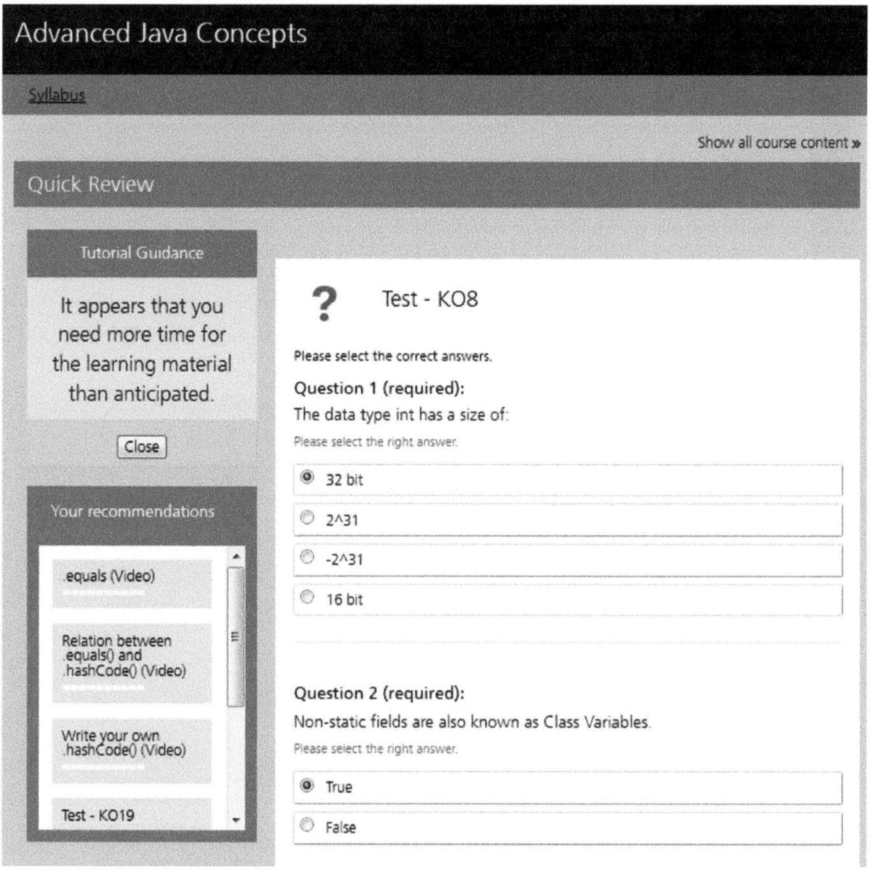

Figure 4.23 Do a test at the end of a Concept Container.

In the standard view of an online course in the IMC Learning Portal the user can also see which Knowledge Object he or she has already successfully passed. This complements the guidance given by INTUITEL.

4.4.2 Network Design

Elena Verdú, María J. Verdú, Luisa M. Regueras, Juan P. de Castro

The University of Valladolid (UVa) has developed a Cognitive Model for the Knowledge Domain or KD "Network Design". This KD is defined for the course "Laboratory of Design and Configuration of Networks", which is offered at the Telecommunications School of the University of Valladolid. The course is a part of the third year curriculum (out of four) of one of the official degrees given at that School: "Degree on Telecommunications Specific Technologies, mention in Telematics Engineering", and it is supported by Moodle through the Virtual Campus of the UVa.

The course "Laboratory of Design and Configuration of Networks" comprises 15 seminar hours (one one-hour session per week) and 45 laboratory hours (one three-hour session per week) during the 15 week long semester. It does not include any frontal lecture while it is an eminently practical subject, where concepts about communication networks already studied in previous courses are reviewed and applied in order to be able to design and configure communication networks.

Specifically, the course adapted to INTUITEL consists of twenty-four lessons or CCs (Concept Containers) about different aspects related to simulation, network design, IP networking, Wide Area Networks (WAN), Local Area Networks (LAN) and Structured Cabling Systems (SCS). These twenty-four CCs include all the learning objects or KOs (Knowledge Objects) of the Moodle course (both the resources as the activities and communication tools) and are organized through two Macro Learning Pathways as listed in Table 4.3: "Classical Learning Path" and "Alternative Hierarchical Learning

Table 4.3   Macro Learning Pathways for the KD "Network Design"

LP Title	Description
Classical Learning Path	This Learning Path guides you logically through topic clusters as it is usually done in the typical literature of this domain: different types of networks are studied sequentially.
Alternative Hierarchical Learning Path	This Learning Path guides you hierarchically through topic clusters, classifying them by topics and then by type of network.

Path". The first one follows an approach near to "Chronologically from old to new" pathway, while the second one follows a "Hierarchically top down" pathway. The Table 1 shows the defined relations.

Figures 4.24 and 4.25 show the CCs belonging to the KD "Network Design" as well as the relations between them in order to form the sequence of CCs for both macro Learning Pathways: "Classical Learning Path" (Figure 4.24) and "Alternative Hierarchical Learning Path" (Figure 4.25). Gray boxes in Figure 4.24 are used for CCs that belong exclusively to the Classical Learning Path. In Figure 4.25 gray boxes are used for CCs that belong exclusively to the Alternative Hierarchical Learning Path. In both figures white boxes represent CCs that are reused in both macro Learning Pathways (shared CCs).

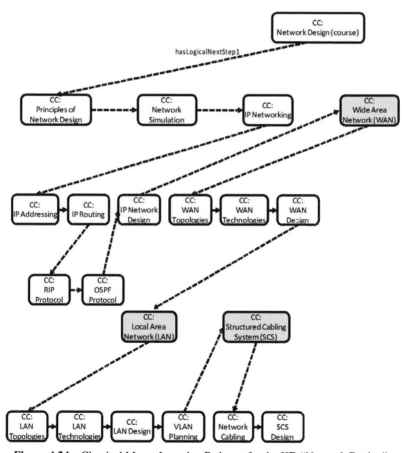

Figure 4.24 Classical Macro Learning Pathway for the KD "Network Design".

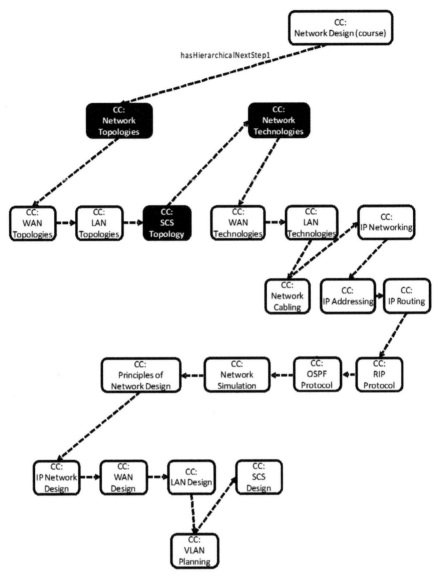

Figure 4.25   Alternative Hierarchical Macro Learning Pathway for the KD "Network Design".

Within each CC, Micro Learning Pathways are used to connect KOs. As a good network design requires examples and practice, the Good-Practice Multi-Stage Learning Pathway has been the most implemented Micro Learning

Pathway, but others have also been used (Simulated Multi-Stage Learning Pathway and Structured Inquiry-Based Learning Pathway). So, for example, the CC "Network Cabling" supports three different Micro Learning Pathways: Good-Practice Multi-Stage Learning Pathway with two different media types (text and video) and Structured Inquiry-Based Learning Pathway (dashed-line arrows and solid-line arrows in Figure 4.26, respectively).

By the dashed-line learning pathway, students receive an explanation of the different types of network cabling through a presentation or an explicative video; while by the solid-line pathway, several questions about the different types of network cabling are presented to students in order to be themselves who learn, study and present the results to their colleagues. Later, in both cases, the students have a tutorial step-to-step (with text or with video) and finally, they must do an exercise about network cabling.

Finally, Table 4.4 shows some examples of the behavior of the INTUITEL intelligent tutoring system depending on the behavior expected by the students

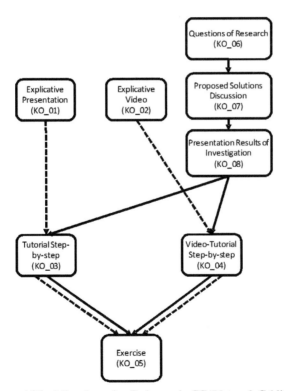

Figure 4.26 Micro Learning Pathways in CC "Network Cabling".

Table 4.4   Examples of the INTUITEL behavior for the CC "Network Cabling"

KO	Learner Behavior	INTUITEL Behavior
KO_01	Read the slides during 7 minutes	Recommend KO_03 as next KO
KO_01	Read the slides during 1 minute	Recommend read again KO_01 with a message indicating that it is necessary to read again this KO because 1 minute is an insufficient time.
KO_02	View the video during 5 minutes	Recommend KO_04 as next KO
KO_04	View the video during 7 minutes	Recommend KO_05 as next KO
KO_05	Do the exercise and answer incorrectly the questions proposed	Check if the student prefers text or video: In case the user chooses "Text" recommend KO_03 as next KO with a message indicating that it is necessary to read again the KO_03 in order to review the concepts. In case the user chooses "Video" recommend KO_04 as next KO with a message indicating that it is necessary to view again the KO_04 in order to review the concepts.
KO_05	Do the exercise and answer correctly the questions proposed	Check the Macro Learning Path followed by the student. If this is the "Classical Path" Recommend the first KO of CC "SCS Design" (see Figure 4.24). If it is the "Alternative Hierarchical path" recommend the first KO of CC "IP Networking" (see Figure 4.25).

for the CC "Network Cabling" shown in Figure 4.26. The table uses the short identifiers of the knowledge objects. For their according titles see Figure 4.26.

4.4.3 Special Relativity

Peter A. Henning

The concept of Learning Pathways (LP) and of navigation in the space of learning objects is most easily demonstrated by using a very simple course exhibiting only two Macro and two Micro Learning Pathways (MLP and mLP). For this simple example we have chosen a course on Special Relativity that we ran on an ILIAS platform.

Special Relativity has been developed in 1905 by Albert Einstein, first of all as a mathematical model to explain certain experimental findings in physics. However, as it turned out, the equations of Special Relativity had a very high predictive power they allowed predictions far beyond those simple experiments. Since then, Special Relativity has proven to be a consistent model of space and time under certain (=special) conditions, and is nowadays believed to be the correct mathematical description of nature's reality under these conditions. In the language of natural science, such a mathematical description is called a *theory*. Note, that in many other fields of knowledge the term "theory" is used in a much weaker sense, e.g. as a kind of personal hypothesis that might (or might not) be true. While we will adhere to the stronger meaning of *theory*, nevertheless the wording shall be avoided in the following.

The role of Special Relativity (henceforth abbreviated as SR) as model of space and time has fostered great interest even from outside the physics community as soon as it was published. This is due to the fact, that even the most basic models of philosophy require a model of space and time, as was already known to the ancient Greek philosophers (and, most probably also to thinkers in pre-historic times). Indeed, apart from abstract considerations like those presented by Immanuel Kant about language, thinking and being (and linked to the concept of absolute space and absolute time), even very concrete socio-political considerations require such a model of space and time. How could one have a concept of "neighbor" without a notion of "distance"? How could one have a concept of "causality" without a notion of temporal "before" and "after"?

Obviously, the domain of Special Relativity bears interest in Physics as well as in Non-Physics and these two views therefore will be used as our Macro Learning Pathways in an adaptive e-Learning course. Let us also mention, that these two views on our domain are semantically very different. In this case, our idea of Macro Learning Pathways exhibits a strong similarity to the Semantic Views Model, and to recent semantic clarifications of the role of theories in science [4]. It must be stressed that in most practical applications of adaptive learning systems such a clear distinction of Macro Learning Pathways will hardly be possible. On the other hand, the distinction according to the target group may serve as a blueprint for the application of the INTUITEL technology in an industrial setting, where the learning content is structured according to certain business roles. Within each of these two MLPs, we will present only five Concept Containers (CC).

In the Non-Physics MLP, an introductive CC will be followed by

- CC "nineteenth century", depicting the space-time model and experimental situation at the end of the nineteenth century.
- CC "twentieth century", depicting the findings in the period 1900–1905.
- CC "Annus Mirabilis", describing in detail the wondrous success achieved by Albert Einstein in is miracle year 1905 from a non-physicists view and a conclusive CC.

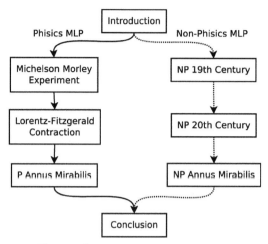

Figure 4.27 Macro Learning Pathways.

In the Physics MLP, the same introductive CC will be followed by

- CC "Michelson-Morley Experiment" hinting towards the non-existence of a ether.
- CC "Lorentz-Fitzgerald Contraction providing a solution to the dilemma that was then taken up by Einstein.
- CC "Annus Mirabilis", describing in detail the wondrous success achieved by Albert Einstein in is miracle year 1905 from a non-physicists view.

Finally, both the Physics MLP and the Non-Physics MLP share a conclusive CC. Figure 4.27 gives an overview of the two MLPs.

As the next step, we will have to specify the two micro learning pathways. The term "micro" was deliberately chosen for this ordering principle, because it is intended to be changed locally in a learners' traversal of the space of learning objects. Note, that this does not mean "instantaneously", as the progress of the current learning object might require to finish it, to finish even a short segment of a given mLP before changing to another. A choice of mLP therefore can be achieved by reasoning on the basis of the current Learner State Ontology including the pedagogical factors in all their variety:

- Learner properties, like age, gender, culture
- Learner behavior
- Learner success
- Dialogue components

- Soft factors like stress and motivation
- Hard factors like bandwidth and environmental information

For our simple example course we will restrict ourselves to a single hard factor determining the mLP: The available bandwidth from the server to the learners' digital interaction device. In particular, we will differentiate only a high-bandwidth and a low-bandwidth mLP. This implies that the media type presented to the learner is chosen according to the bandwidth. However, the INTUITEL reasoning process is not reduced to selection of a proper media, because a proper annotation of the learning objects may still indicate to the learner, that some LO are more adequate than others. This will be outlined below.

The simple bandwidth mLP are depicted in Figure 4.28 for one of our CC. The dashed frames indicate the first and the last element on a mLP. In this CC we have only three Learning Objects:

- A summarizing text, present on both mLP
- A concluding text, present on both mLP
- A virtual experiment, consisting of an interactive video clip and therefore only present on the high bandwidth mLP

Choosing the low bandwidth mLP for the concluding CC therefore only means: Leave out the virtual experiment. Now let us assume, that a learner enters the course with the selection of non-physics MLP and high-bandwidth mLP. This selection may either be done manually or may be determined from information gathered about the learner, like e.g. previous choices, a simple question asked at the beginning of the course and a bandwidth detection component.

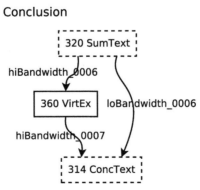

Figure 4.28 mLP for the Concept Container "Conclusion". Dashed frames indicate the first and the last element of a mLP.

On entering the course, the learner will be shown the KO Course Overview usually a short page of meta data, indicating how much time one would need etc. Simultaneously to a configurable sound message, a selection box (see Figures 4.29 and 4.30) will appear at the side of the screen (depending on configuration) that presents to the learner the following three KOs along his current mLP however, not in the ordering they have within the concept container. Instead, the highest recommendation (four stars) is given to the last KO of this CC the KO titled "History 19th century".

The reason for this is the additional meta data for the course: It attributes a low learner knowledge level to the other two KOs and a high learner knowledge level to the "History" KO. Since the learner also comes with additional meta data, attributing him to be an experienced learner (or, at least one who has visited the course already three times), the INTUITEL reasoning process determines that it would be more suitable for our particular learner to proceed directly to the most difficult KO in this CC. Note, that nevertheless full flexibility is maintained: The learner may also choose any other KO in the current CC and will then be recommended the remaining KO along his mLP.

Now, what happens if the learner follows the first recommendation, and proceeds to the "History" KO? Of course, he is shown the corresponding KO but at the same time receives already a recommendation for the following KO (see Figures 4.31 and 4.32). These are the KOs of the CC along his current MLP (Non-Physics), in this case only two KOs with equal recommendation level.

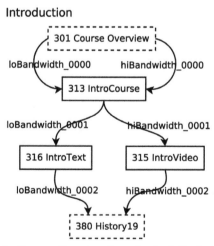

Figure 4.29 mLPS in the Concept Container "Introduction". Dashed frames indicate the first and the last element of a mLP.

Figure 4.30 The selection box shown to the learner when entering the Concept Container "Introduction".

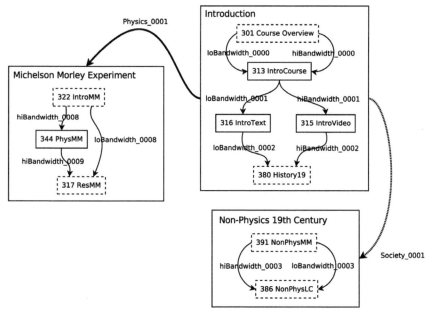

Figure 4.31 Branching of the MLP on exiting the first CC. Dashed frames indicate the first and the last element of a mLP.

However, it is easily possible to set up the INTUITEL system such that from the last KO in the current CC one is also shown the first KO from a CC on another MLP. Consequently, in such a setup the learner will be given the chance to change the Macro Learning Pathway.

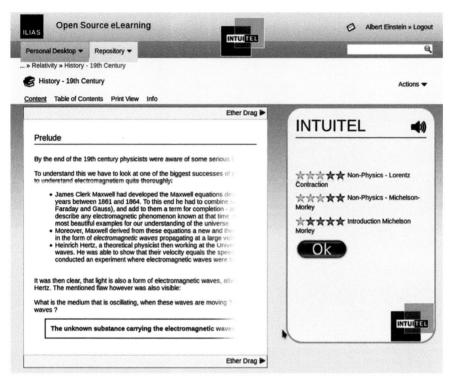

Figure 4.32   Selection box shown in parallel to the last KO of the first CC. We present here a screenshot from the ILIAS implementation of INTUITEL, with the "intuitel/integrated" ILIAS skin.

Currently, the INTUITEL system does not allow leaving the current MLP while somewhere inside a CC one always has to proceed to the next branching point. The modular and transparent concept of the INTUITEL software allows changing this behavior with a few lines of Java code but as a result of the INTUITEL project we have understood the pedagogical reasoning process to a great detail and would suggest a different approach: If a learner is motivated to change the MLP while deep inside some CC, the logical structure of the course should be reconsidered, and possibly some more common CCs should be used.

Figure 4.33 depicts the overall structure of the INTUITEL course on Special Relativity. We will not elaborate on the other KO and CC, since in this simple design each MLP has to be followed after it has been chosen the two MLP come together again in the concluding CC.

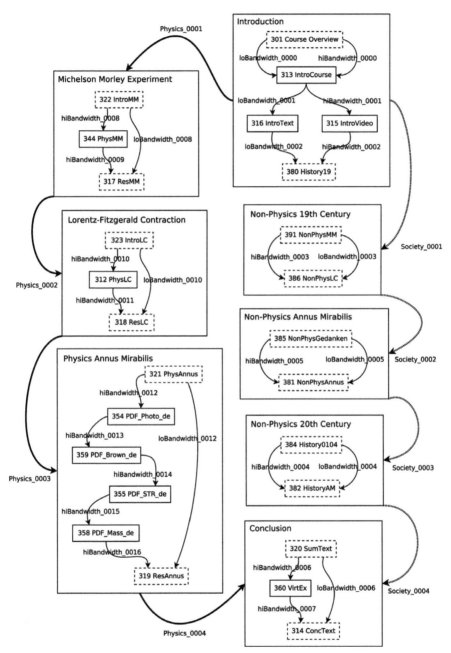

Figure 4.33 Overall structure of the INTUITEL course on Special Relativity. Dashed frames indicate the first and the last element of a mLP.

With only two MLP, two mLP and 26 KOs it is a rather simple example for the potential of INTUITEL. And yet, it makes clear that the main task now lies at the hands of the cognitive engineer: Designing the proper course structure and entering the necessary meta data is the key to successful adaptive learning.

4.5 Evaluation and Testing

Luis de-la-Fuente-Valentin and Daniel Burgos

Frameworks for Technology Enhanced Learning are really difficult to evaluate because of the large amount of factors that may affect the learning scenario. Let's say that a bad design of the user interface hinders a student's interaction with the system. In that case, the student's perception won't be as positive as expected, regardless of how useful the recommendations are. Therefore, the validation has to be very careful with the aspect being tested.

From a technical perspective, the framework development was based on a distributed unit-test approach, in which unit-tests were centralized in a single server and deployed into the different INTUITEL modules as REST requests. This centralized process, daily triggered, guaranteed that all the distributed software modules were compliant with the Communication Layer that mediated all the request-response messages in the framework.

From a pedagogical perspective, the validation was carefully designed to support a set of predefined test cases. The support for these test cases was validated by introducing the concept of "artificial test learners", generated by collecting typical data from institutions of higher education and from the commercial educational interest companies. These test cases are available as an internal working document and are then fed into the INTUITEL enabled prototypes. This section depicts and details the above described validation techniques, focusing on them as a method to guarantee the required reliability prior to the deployment of the framework in a real learning scenario.

Tests on the Communication Layer

The INTUITEL software architecture depicted in 4.4 uses a centralized strategy for messaging in which the Communication Layer receives and delivers every message exchanged in the Recommendation Process. This strategy allows for a distributed software architecture where the different modules may run in different servers as long as they know the location of the Communication Layer. Therefore, the Communication Layer is a key module in the INTUITEL system and must be carefully tested.

A unit-test based testing strategy has been followed. The testing suite was developed during the first months of the project and ensured the homogeneity of the REST messages structure. The testing was executed as follows:

1. For each component being tested, correct and incorrect (designed on purpose) REST messages were sent.
2. The testing suite validated if the obtained response matches the expectations, according to the Communication Layer specifications and the USE/TUG/LORE message definition (see Section 3.5).
3. A report was built and published including the result of every test case.
4. Whenever a mistake was found on a response message, the report was emailed to the developer of the corresponding partner.
5. This process was daily repeated.

It can be noticed that this strategy treated the tested servers as passive subjects, where they only have to offer an endpoint to provide successful answers to the REST messages they receive. However, each server may also send messages in a proactive way. In such case, the Communication Layer was in charge of verifying the correctness of the different REST messages.

As explained in Section 3.5 and provided in Table 3.3, INTUITEL defines eight endpoints for REST messages: lmsprofile, learners, login, mapping, TUG, LORE, USE/performance and USE/environment. For each of those endpoints, the testing suite composed and sent a total of seven different messages:

1. Correct message with valid XML, correct values in HTTP headers and namespaces properly set.
2. A message with valid XML, correct values in HTTP headers but no namespaces defined.
3. A message with valid XML, but incorrect values in HTTP headers.
4. A message with invalid XML (bad element name)
5. A message with invalid XML (bad attribute name)
6. A message with invalid XML (empty attribute value)
7. A non-XML message

The correctness of the response was decided by verifying the HTTP Status Code of the response. A 200 OK status code was expected for messages 1 and 2, while 4XX was expected for the rest of the test cases. The test suite used the following framework:

- REST messages were generated from static XML files and sent via PHP CURL libraries.

- The PHP-Unit library is used to organize the test messages as unit tests and generate XML structured output.
- The PHING Framework is used to generate HTML human readable reports from the XML structured output.
- Doxygen is used to generate source code documentation, so that every report includes the test results and the description of the test cases.
- The test suite scheduled with cron to be executed once a day, and reports were sent to the developers with the UNIX send mail utility.

The testing suite does not verify if the system generates recommendations. It just ensures that all the different modules are alive and compatible with the Communication Layer. The correctness of the provided recommendations are verified by the so called artificial test learners.

Artificial Test Learners: Technical Tests on the INTUITEL Recommendations

The artificial test learners were implemented as software pieces that executed actions similar to those expected from actual learners. Such actions fed the LPM and the INTUITEL Engine so the test cases checked if the produced recommendation was correct according to the INTUITEL expected reasoning. The overall behavior of the testing is depicted in Figure 4.34. Other tasks in the project were devoted to test the LPM and the INTUITEL Engine:

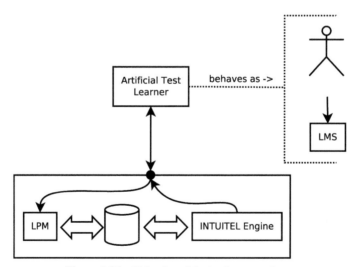

Figure 4.34   Behavior of the testing procedure.

- LPM testing was fed with student actions and examined the selected Learning Pathway. The LPM calculates the learner's current position, and determines the most suitable Learning Pathway based on the history of the learner. The LPM does not produce any recommendation, so LPM testing did not test recommendations.
- INTUITEL Engine testing was fed with LPM output, and the unit testing was devoted to check if the recommendation is the expected one.

Therefore, the two described subsystems do not test the correct behavior of the overall system. Artificial test learners executed the testing from the perspective of the learner: feeds the system with the user actions, and checks if the recommendation (i.e. the communication with the learner) is the expected one.

Changes in the Initial Approach

The initial idea of the artificial test cases (and therefore described in Deliverable 12.1 [24]) was to apply a LMS based approach. That is, the artificial test learner was conceived as an automated browsing (web scrapping) through the course contents, programmed with web scraping techniques such as mechanize, casper or any other similar software library. However, further discussions within the topic revealed the following problems for such approach:

- Testing would be LMS specific. That is, test cases should be re-written for each of the INTUITEL enhanced LMS.
- Timing cannot be efficiently reproduced. That is, if the student takes 2 hours before choosing a KO, then the automated script would take the same 2 hours.
- Problems in the LMS would distract from the actual subject being tested: LPM+Engine.

Therefore, the researchers decided to change the approach and design the artificial test learners as described in this document.

Description of the Test Cases Template

A template guided the development of the test cases used to test the system. This template consists in two main sections: first the description of the initial state, including the values of all Didactic Factors and any other information stored in the database that could be useful to determine the learner state.

Taking this description of the initial state as the starting point, the second part of the template includes the step-by-step students' actions. That is, the messages that the LMS would send to the LPM as a response of the student's actions within the course. More specifically, the template to specify test scenarios is defined by:

- *Initial State Description*: Preliminary data describing the scenario and learner mostly based on the LPM ontology (Didactic Factors and values in Deliverable 3.2 [25]). Didactic Factors can be taken in and out according to the scenario and test requirements. Another important element of the initial description is the database snapshot that describes the student's previous activity. In other words, a list of which learner has accessed which KOs in which order.
- *Student actions*: A sequence of learner's actions. That is, a temporarily ordered sequence of LSO and "reflex" input messages. The test case also included the temporarily ordered sequence of LORE and TUG output messages, so the artificial test learners could check if the received message is the expected one.
- *SLOM metadata*: CM and CCM are needed for reasoning, so SLOM meta data should be included as part of the test case description.

The template for the test cases included the technical information required for the development of such a technical development as the artificial test learners. However, some textual description of the test case is also needed for the proper understanding of the scenario and, if possible, for the reuse of the test cases in other testing tasks.

Data Required to Represent a Test Case

The artificial test learners fed the LPM with the events produced by the learner at the LMS, and verified the result from the INTUITEL Engine. Therefore, the first data required is the piece of information sent from the LMS to the LPM at a specific situation. Second, the LPM will produce an output that will inform the Engine of the most suitable LP and other relevant information, which is the LSO. Third, the Engine will produce an output to be sent back to the LMS, shaped as USE/TUG/LORE messages. Finally, the SLOM data that defines the content itself is required for the testing in order to contextualize the expected output data. In summary, the required pieces of information for the test cases are:

- The messages sent by the LMS: Learner Update and TUG/LORE responses.
- SLOM related to the content needed for reasoning.
- Corresponding snapshots of user database or model needed for reasoning.
- LSO and "reflex" messages produced by the LPM.
- Intuitel Engine responses, shaped as USE/TUG/LORE messages.

5

Conclusion and Outlook

Kevin Fuchs

In this final chapter, we briefly summarize the INTUITEL project and the novelties it brings to the field of adaptive learning environments and learning analytics. INTUITEL itself is a system that operates on didactic knowledge and meta-knowledge respectively. However, it is not a tool for retrieving such knowledge. Therefore, in this section we give an outlook to current and future work to close this gap.

5.1 Summarizing INTUITEL

Adaptive learning environments seek to adapt the selection and presentation of learning content with respect to the learner's individual characteristics. Amongst many more, this includes social, economic, cultural and ethical background, gender, physiological and psychological abilities and disabilities, the learning environment, the technical devices used, the learner's prerequisite knowledge, her learning progress and physical or emotional states.

For example the impact of arousal and stress has been well explored and already utilized for the development of affective game designs [3, 84, 86, 106]. In the psychological field, there have been multiple suggestions to define learning styles and partly also how to derive appropriate course designs from them [43, 54, 55, 57, 68]. Large scale studies across Europe have disclosed significant correlations of learning success on the one side and factors like gender, cultural, ethnic, and socio-economic backgrounds on the other side [44, 45]. It is hence a worthwhile goal to enrich technology-enhanced learning with the capacity of detecting the above described factors and including them into decisions on how to guide a learner.

The INTUITEL system provides a novel and innovative way of transforming human-based didactic expertise into a formal, machine-processable representation. By using ontologies for the representation of both didactic knowledge and Didactic Factors we pattern human language and thinking.

This way, no particular technical expertise is needed in order to create meaningful content. INTUITEL decouples technology from didactics entirely. The annotation of learning material is, therefore, intuitive and applicable to any domain of knowledge.

The result of INTUITEL-made deduction processes is a virtual tutor within a common Learning Management System like ILIAS or Moodle. However, INTUITEL does not enforce certain learning pathways to the learner. Instead INTUITEL only gives recommendations to the learner without obligation. The hypercube model and the calculation of cognitive distances between the learner's state and recommended, predefined learning pathways makes INTUITEL constantly generate new recommendations regarding the learners actions.

As a theoretical foundation we may formulate four universal criteria a learning environment has to satisfy to be adaptive with respect to learning style, behavior and preferences of individual learners:

1. Indicators have to be found to detect factors of interest. These indicators have to be measured by or sent to the system constituting the adaptive learning environment.
2. The indicators must be aggregated to the desired factors and there must be a formal machine-processable representation of them.
3. In the context of the aforesaid formal representation, didactics and learning content must be associated with those factors that are supposed to have impact on learning behavior.
4. The learning environment must then deduce appropriate learner guidance from this formal representation.

5.2 Operationalization of Didactic Factors

With the INTUITEL project we developed an approach that is able to fulfill these requirements. However, the remaining problem is the identification and operationalization of influencing factors. This challenge is mostly located in the domain of social and educational science and forms a preliminary condition for the design of adaptive learning environments.

The INTUITEL system itself is capable of using such information for its recommendation process. Yet this information has to be provided for the system. INTUITEL does not generate the according data by itself. Therefore as a future task, we will introduce an approach on how to find and utilize influencing factors to use within an adaptive learning environment like INTUITEL.

Here, we particularly address time as an immanent dimension of learning. Time has a special meaning for learning in two aspects: first learning always is a transformation of semantically connected content into a linear sequence along the time dimension. Second, many of the factors we supposed to have an impact on learning behavior and success are time-dependent functions. In a novel way, the following approach models learning histories of individual learners as spatio-temporal trajectories using techniques from the field of spatio-temporal databases. By this, we aim to provide a new way of learning analytics.

Let us briefly reflect on the foundations of INTUITEL. The formal representation of learning content, influencing factors and didactics is implemented by ontologies from which the INTUITEL system deduces learning recommendations. The set of these ontologies comprises the following parts:

- A pedagogic ontology based on the web didactics by Norbert Meder. This ontology organizes the learning content by courses, units, and atomic knowledge objects (KOs) [61, 65].
- A learner state ontology, describing the learner's current state, progress as well as personal and environmental characteristics [25]. This ontology is also the part in which information about influencing factors is included.

In INTUITEL so called "Didactic Factors" as explained in Section 3.2 are of central meaning. They are identical with the influencing factors described previously. A Didactic Factor is aggregated by measuring one or more indicators. Indicators may represent non-nominal data. By particular transformation rules defined for each factor this non-nominal data is transformed into nominal data. This nominal data forms an identifier denoting an individual that is part of the INTUITEL ontologies. This way the factor exists in a formal and machine-processable form that we claimed before.

In order to calculate the learner's position within the learning environment, INTUITEL uses the hypercube model as we elaborated in Section 3.3. Remember that each of the n dimensions of that hypercube represents a KO. Each of these dimensions is assigned a numeric value representing the state of progress a learner has performed on the according KO at a certain point in time expressed by a value from the interval [0, 1]. A learner's position in this space is a vector $L = (l_1, \ldots, l_n)$ evolving over time and thus drawing a trajectory in the n-dimensional space. The trajectory of any vector L is then located within that hypercube. Such trajectories must not be mistaken for the "learning pathways" that INTUITEL uses (see Section 3.3). Those learning pathways only consist of linear sequences of KOs with no explicit relation to the time dimension. However, a trajectory in the hypercube represents the entire learning history of an individual learner.

5.3 The Hypercube Database Project

The fact that a learner's position is projected to a trajectory over time has not yet been utilized by INTUITEL nor has it in other systems known to the authors. Indeed, this is subjected by current and future work summarized under the "Hypercube Database" project [32].

INTUITEL provides the technology to use Didactic Factors for adaptivity in learning environments. However, those Didactic Factors still have to be operationalized. While the INTUITEL approach provides a universal process as well as the fundamental technology to associate Didactic Factors with learning content, it remains a challenging question how such factors are supposed to be identified and how they can be measured by sensor data.

The Hypercube Database project aims to combine learning analytics with the technology of spatio-temporal databases. Learning pathways of individual learners together with arbitrary indicators influencing learning behavior are measured and stored with a spatio-temporal database in the form of high-dimensional trajectories interpolated over the time dimension. For this purpose, we enhance the hypercube model of INTUITEL to utilize the trajectories of the learners' positions together with measured sensor data that contribute to Didactic Factors.

5.3.1 The Advanced Hypercube Model

The hypercube model is enhanced by k additional dimensions. This way the model describes an $(n + k)$-dimensional space with n being the number of *KOs* in a learning environment and k being the number of measured indicators. Each of the k dimensions that stand for indicators is assigned a numeric value representing the value that is measured for this indicator at a particular point in time. A learner's position in this space is now a time-dependent vector $M = (m_1, \ldots, m_n, m_1 \ldots m_k)$ forming a trajectory in the $(n+k)$-dimensional space. The k dimensions may be normalized, which is not necessarily required.

Learners' movements through this $(n + k)$-dimensional space define trajectories which we will model by the use of a spatio-temporal database that is – at the publication date of this book – yet in the process of development. The basic idea is to perform cluster-analysis solely on the basis of geometric relations between the trajectories of multiple learners in order to identify common learning pathways and learner groups. In subsequent analysis, we want to find out, which indicators correlate to these pathways in order to predict the learning behavior of individual learners.

The major difference to common data analysis is the fact that all information – including the k indicators – is transformed into purely geometric information and thus lifted to a highly abstract level. We only consider the geometric properties of and geometric relations between hyperpolylines. Arbitrary dimensions may or may not be included into analysis by simply performing projection. The model is entirely open to add and remove any kind of variables and dimensions respectively as long as they are numeric.

5.3.2 Example Applications

We now sketch two examples to illustrate how such data analysis can contribute to the improvement of adaptive learning environments.

i. Discovery of Unknown Didactic Factors: Within an experimental learning situation, arbitrary indicators are measured. The resulting data is transferred into the above described system and the data is converted into persistent learning histories together with their indicators. Using factor analysis, new Didactic Factors can be identified together with indicators that are represented by these factors. In a second step, the original set of indicators can be reduced to a smaller one, restricted to indicators that are easy to measure in a non-experimental learning environment.

ii. Real-Time Learning Pathway Prediction: Like in the previous example, learning histories as well as influencing indicators are stored with the advanced hypercube model. In a first stage, the learning histories are subjected to a cluster analysis in order to identify common classes of learning histories. In the second stage, taking these clusters on the one side and the measured indicators on the other side, one can perform for example either a discriminant analysis or a logistical regression. As a result, we can determine which variation of indicators of a specific learner will probably lead to a specific learning history. Built on this knowledge and measuring these indicators in the learning environment, i.e. in an LMS, we can predict the learner's future learning history and recommend according learning pathways and KOs.

5.3.3 Implementation of the Hypercube Database

There are various approaches of existing spatial, temporal and spatio-temporal databases. Purely temporal databases are for example the ARCADIA database for clinical applications [19], Calanda for time series with financial data [89], ChronoLog running on top of a standard Oracle database [8], HDBMS [18],

TDBMS [101] and TimeDB for general purpose which is based on the ATSQL2 query language [15, 16, 95, 100].

The field of spatio-temporal databases is mostly dominated by Geographical Information Systems (GIS), Network and Facility Management, Land Information Systems (LIS) and Image Processing [1]. For example GRASS GIS [70] and GeoToolKit [5] are Geographical Information Systems while the CONCERT database focuses on management of raster images [81, 82]. The SECONDO database – developed at the University of Hagen – is a multipurpose system for spatio-temporal data [41, 105]. Due to the nature of their subject these systems mostly provide support for only two or three spatial dimensions. The DEDALE database is capable of dealing with higher dimensions and is based on a constraint database technique [36–39, 83].

All databases dealing only with two or three spatial dimensions are not an option for the Hypercube Database due to its high-dimensional space. The DEDALE system appears to be an interesting candidate because of its constraint approach that can be exploited for any number of dimensions. However, the constraint database approach is most appropriate for querying geometric objects containing infinite point sets whereas it is less suitable for querying continuous trajectories. We will therefore develop our own database but we will use the temporal database TimeDB as its back end and build the spatio-temporal functionality upon it.

At the publication date of this book, the implementation of the Hypercube Database is still in progress. Therefore the following elaborates current and future work. We describe the architecture of the software as well as the data structures and algorithms we intend to use. Figure 5.1 shows the fundamental parts of the system which is subdivided into the Vector Module, the Hypercube Module and the Database Access Module.

The system is written in Java and uses the temporal database TimeDB as a back end which is managed and accessed by the *Database Access Module*. TimeDB itself provides temporal support only for database tuples but not attribute-wise. The *Database Access Module* built upon TimeDB provides an interface for the *Vector Module* with which temporal support for single attributes is achieved.

The task of the *Vector Module* is the transformation of single measuring points (with respect to indicators) into temporal vectors and storing it via the *Database Accessor Module*. Consider an individual (a learner) for which we want to measure the values of m variables over time. For each variable we measure values at arbitrary points in time. For each variable we regard the lastly measured value as valid until a new value is measured. Alternatively,

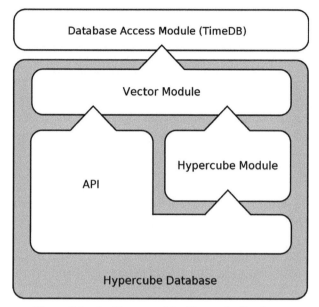

Figure 5.1 Architecture of the Hypercube Database.

we can also interpolate between two measured values. This way, we get an *m*-dimensional time-dependent vector for each individual. The listing below describes the transformation of measuring points into vector representation. We illustrate the subsequent insertion/deletion of measuring points and the evolution of the vectors for a trajectory with three variables a_1, a_2 and a_3. At the beginning all variables have an initial value, e.g., 0 from start to eternity.

$$a_1 = 0 \; for \; t \; \in \; [0, forever)$$
$$a_2 = 0 \; for \; t \; \in \; [0, forever)$$
$$a_3 = 0 \; for \; t \; \in \; [0, forever)$$

Now, we insert the measuring points $a_2 = 3$ at the time point t_1, $a_2 = 5$ at t_2 and $a_1 = 7$ at t_3

$$a_1 = \begin{cases} 0 & for \; t \in [0, t_3) \\ 7 & for \; t \in [t_3, forever) \end{cases}$$

$$a_2 = \begin{cases} 0 & for \; t \in [0, t_1) \\ 3 & for \; t \in [t_1, t_2) \\ 5 & for \; t \in [t_2, forever) \end{cases}$$

$$a_3 = 0 \; for \; t \in [0, forever)$$

The *Hypercube Module* finally is responsible for transforming these vector data into spatio-temporal trajectories as described by the advanced hypercube model. Within the *Hypercube Module* we will implement indexing and querying functionality in order to access the trajectories efficiently and to cluster them by spatio-temporal characteristics.

There are multiple indexing techniques for spatio-temporal data. Many of them are based on the R-Tree family [42] for multidimensional spatial indexing. [40] lists, e.g., the 3D R-tree, the HR-tree, the RT-tree and the MR-tree. Moreover – for indexing moving objects with respect to the current time and the near future – [40] refers to TPR-trees, multilevel partition trees, kinetic B-trees and kinetic external range trees.

The usefulness of those indexing techniques strongly depends on the kind of the data and the kind of queries to be performed. In the case of the Hypercube Database we are less interested in querying point sets like "select the geographic region that was covered by the storm between 5 am and 7 pm". Such a query would be useful within a Geographical Information Systems and would return a point set as a geometric object altering over time. But in our case we are mostly interested in queries referring to entire trajectories like "select all trajectories close to trajectory *x*".

Appropriate methods for indexing and querying trajectories are for example the Spatio-Temporal R-tree (STR-tree) and the Trajectory Bundle Tree (TB-tree). Both index structures are appropriate for performing point, range and nearest-neighbor queries as well as trajectory-based queries [40]. However, using such indexing techniques like the STR-tree or the TB-tree are problematic for the high dimensional hypercube space. In particular, the performance of indexing rapidly decreases with the growing dimensionality. Addressing this particular issue, the X-tree is another index tree structure that is a variation of the R-tree and that is designed especially for high-dimensional data [7].

Another promising candidate is the grid index [17]. This index structure represents its bounding boxes in the form of static cells that are organized in a grid instead of a search tree. This structure is especially interesting for trajectories as they grow monotonically along the time dimension with few or no modifications after insertions. The grid index treats spatial and temporal indexing separately. This means that for temporal indexing any other method may be used. This way the grid index is a perfect candidate to be combined with the above described temporal index structure based on TimeDB.

5.4 Conclusion

We have defined four universal criteria a learning environment has to satisfy to be adaptive with respect to learning style, behavior and preferences of individual learners. Firstly, Didactic Factors have to be retrieved by measuring correlated indicators. Secondly, these factors have to be transformed into a machine-processable form. Thirdly, the Didactic Factors have to be annotated to learning content, together with didactic relations between pieces of learning content. Fourthly, the learning environment deduces the according instructional design from this formal representation.

INTUITEL satisfies the second, third and fourth of these requirements. With the Hypercube Database project we aim to close the gap to the first requirement, designing and developing a research tool for the analysis of learning histories. We model learning histories as spatio-temporal trajectories treating the time dimension as an immanent part of learning. Besides the learning content itself, the concept of the advanced hypercube also includes arbitrary additional data that may result from measured indicators. By this – inside the space of the advanced hypercube – data is lifted to a highly abstract level, mapped to purely geometric information.

This leads to a compact representation allowing us to analyze a wide range of data solely on the grounds of hyperpolylines, their spatio-temporal characteristics and their relations to each other. Not only is this a new application of a spatio-temporal database. It also offers a new approach for finding common learning pathways and Didactic Factors correlating with them. By this, we can predict learning pathways by observing a learners' current actions and retrieving the according Didactic Factors, which constitutes the enhancement of adaptive learning environments in the future.

References

[1] T. Abraham and J. F. Roddik. Survey of spatio-temporal databases. *GeoInformatica*, 3(1):61–99, 1999.

[2] P. Arapi, N. Moumoutzis, and S. Christodoulakis. Aside: An architecture for supporting interoperability between digital libraries and e-learning applications. In *Sixth International Conference on Advanced Learning Technologies*, pages 257–261, 2006.

[3] P. J. Astor, M. T. P. Adam, K. Schaaff, and C. Weinhardt. Integrating biosignals into information systems: A neurois tool for improving emotion regulation. *Journal of Management Information Systems*, 30(3):247–277, 2014.

[4] J. Azzouni. A new characterization of scientific theories. *Synthese*, 191(13):2993–3008, 2014.

[5] O. Balovnev, M. Breunig, A. B. Cremers, and S. Shumilov. Extending geotoolkit to access distributed spatial data and operations. In *Scientific and Statistical Database Management. 12th International Conference*, 2000.

[6] R. Barchino, J. R. Hilera, L. De-Marcos, J. M. Gutiérrez, S. Otón, J. J. Martinez J. A. Gutiérrez, and L. Jiménez. Interoperability between visual uml design applications and authoring tools for learning design. *Information and Control, International Journal of Innovative Computing*, 8(1):845–865, 2012.

[7] S. Berchtold, D. A. Keim, and H. Kriegel. The x-tree: An index structure for high-dimensional data. In *Proceedings of the Twenty-second International Conference on Very Large Data-Bases; Mumbai (Bombay)*, 1996.

[8] M. Böhlen. *Managing Temporal Knowledge in Deductive Databases*. dissertation, Swiss Federal Institute of Technology Zurich, 1994.

[9] B. Bredeweg and P. Struss. Current topics in qualitative reasoning. *AI Magazine*, 24:13–16, 2003.

[10] P. Brusilovsky. Adaptive hypermedia. *User Modeling and User Adapted Interaction*, 11:87–110, 2001.

[11] B. G. Buchanan and J. Lederberg. The heuristic dendral program for explaining empirical data. In *IFIP Congress*, pages 179–188, 1971.

[12] R. R. Burton. The environment module of intelligent tutoring systems. pages 109–130, 1988.

[13] R. Callois. *Man, Play, and Games*. New York: The Free Press, 1961.

[14] V. Carchiolo and N. Vincenzo L. Alessandro, M. Giuseppe. Adaptive e-learning: An architecture based on prosa p2p network. 4:777–786, 2008.

[15] A. Carvalho, C. Ribeiro, and A. Sousa. Spatial timedb – valid time support in spatial dbms. In *Proceedings of 2nd International Advanced Database Conference—IADC-2006*, 2006.

[16] A. Carvalho, C. Ribeiro, and A. Sousa. A spatio-temporal database system based on timedb and oracle spatial. *Research and Practical Issues of Enterprise Information Systems*, 205:11–20, 2006.

[17] V. P. Chakka, A. Everspaugh, and J. M. Patel. ndexing large trajectory data sets with seti. In *Proc. Conf. Innovative Data Systems Research (CIDR '03)*, 2003.

[18] C. Combi, F. Pinciroli, and G. Cucchi M. Cavallaro. Design of an information system using a historical database management system. In *Proceedings of the 8th. Annual International Conference on Information Systems*, pages 86–96, 1987.

[19] C. Combi, F. Pinciroli, and G. Cucchi M. Cavallaro. Querying temporal clinical databases with different time granularities: the gch-osql language. In *Proceedings of the Annual Symposium on Computer Application in Medical Care*, pages 326–330, 1995.

[20] N. A. Crowder. Teaching machine. us patent number 4043054. www. google.de/patents/US4043054 (30.04.2013), 1977.

[21] J. Dewey. *Demokratie und Erziehung. Eine Einleitung in die philosophische Pädagogik*. Weinheim: Beltz, 2000.

[22] M. Doorten, B. Giesbers, J. Janssen, J. Daniëls, and E. J. R. Koper. Transforming existing content into reusable learning objects. pages 116–127, 2004.

[23] E. Duwal. Attention please! learning analytics for visualization and recommendation. In *In Proc. LAK'11, Banff, AB, Canada*, 2011.

[24] A. Schmoelz (editor), C. Swertz (editor), and A. Forstner (editor). Intuitel – deliverable 12.1: Overall pedagogical testing plan. INTUITEL Resources, retrieved Dec. 11 2015 from http://www.intuitel.eu/resources, 2013.

[25] A. Streicher (editor), F. Heberle (editor), and B. Bargel (editor). Intuitel – deliverable 3.2: Specification of the learning progress model. INTUITEL Resources, retrieved Dec. 11 2015 from http://www.intuitel.eu/ resources, 2013.

[26] E. A. Feigenbaum (editor). *The Handbook of Artificial intelligence*. Los Altos/California: William Kaufmann Inc., 1981.

[27] O. G. Perales (editor) and L. de la Fuente Valentín (editor). Intuitel – deliverable 3.3: Lpm communication standard. INTUITEL Resources, retrieved Dec. 11 2015 from http://www.intuitel.eu/resources, 2013.

[28] P. A. Henning (editor) and F. Heberle (editor). Intuitel – deliverable 1.1: Data model and xml schema for use/tug/lore. INTUITEL Resources, retrieved Dec. 11 2015 from http://www.intuitel.eu/resources, 2013.

[29] P. A. Henning (editor) and F. Heberle (editor). Intuitel – deliverable 4.1: Specification of slom – semantic learning object model. INTUITEL Resources, retrieved Dec. 11 2015 from http://www.intuitel.eu/ resources, 2013.

[30] W. Corell (editor). Braunschweig: Westermann.

[31] H. A.Witkin et al. *Personality through perception.* New York: Harper, 1954.

[32] K. Fuchs, P. A. Henning, and M. Hartmann. Intuitel and the hypercube model – developing adaptive learning environments. *Journal on Systemics, Cybernetics and Informatics: JSCI,* 14(3):7–11, 2016.

[33] J. Giesinger. Bildsamkeit und bestimmung. kritische anmerkungen zur allgemeinen pädagogik dietrich benners. *Zeitschrift für Pädagogik,* 57(6):894–910, 2011.

[34] A. C. Graesser. Learning, thinking, and emoting with discourse technologies. *American Psychologist,* pages 746–757, 2011.

[35] T. R. Gruber. Towards principles for the design of ontologies used for knowledge sharing. *International Journal of Human – computer Studies,* 43:907–928, 1995.

[36] S. Grumbach, P. Rigaux, M. Scholl, and L. Segoufin. Dedale, a spatial constraint database. *DBPL,* pages 38–59, 1997.

[37] S. Grumbach, P. Rigaux, and L. Segoufin. Modeling and querying interpolated spatial data. In *Proceedings 15èmes Journées Bases de Données Avancées, BDA,* pages 469–487, 1999.

[38] S. Grumbach, P. Rigaux, and L. Segoufin. On the orthographic dimension of constraint databases. *ICDT,* pages 199–216, 1999.

[39] S. Grumbach, L. Segoufin, and P. Rigaux. Efficient multi-dimensional data handling in constraint databases. *BDA,* 1998.

[40] R. H. Güting and M. Schneider. *Moving Objects Databases.* Morgan Kaufmann Publishers, 2005.

[41] R. H. Güting, T. Behr, and C. Düntgen. Secondo: A platform for moving objects database research and for publishing and integrating research implementations. *IEEE Data Engineering Bulletin 33:2,* 3:56–63, 2010.

[42] A. Guttmann. R-trees: a dynamic index structure for spatial searching. In *SIGMOD '84 Proceedings of the 1984 ACM SIGMOD international conference on Management of data,* pages 47–57, 1984.

[43] P. Honey and A. Mumford. *The Manual of Learning Styles.* Peter Honey Publications, 1982.

[44] P. Honey and A. Mumford. *PISA 2012 Results: Excellence through Equity (Volume II) Giving Every Student the Chance to Succeed.* OECD, 2013.

[45] P. Honey and A. Mumford. *PISA 2012 Results: What Students Know and Can Do (Volume I, Revised edition) Student Performance In Mathematics, Reading and Science.* OECD, 2014.

[46] R. Hönigswald. *Über die Grundlagen der Pädagogik. 2. umgearb. Auflage*. München: E. Reinhardt, 1927.

[47] I.-H. Hsiao, S. Sosnovsky, and P. Brusilovsky. Guiding students to the right questions: adaptive navigation support in an e-learning system for java programming. *Journal of Computer Assisted Learning*, 12(4): 270–283, 2010.

[48] H. A. Innis. *The Bias of Communication*. Toronto: Univerity of Toronto Press, 1951.

[49] D. H. Jonassen and B. L. Grabowski. *Handbook of Individual Differences, Learning and Instruction*. New York/London: Routledge, 1993.

[50] D. H. Jonassen and B. L. Grabowski. *Visible Learning*. New York: Routledge, 2008.

[51] P. Karampiperis and D. Sampson. Towards a common graphical language for learning flows: Transforming bpel to ims learning design level a representations. In *Seventh IEEE International Conference on Advanced Learning Technologies ICALT*, pages 18–20, 2007.

[52] M. Kerres and C. de Witt. Quo vadis mediendidaktik. zur theoretischen fundierung von mediendidaktik. *Medienpädagogik*, 2, 2002.

[53] J. Klauer and D. Leutner. Weinheim, Basel: Beltz.

[54] A. Y. Kolb and D. Kolb. The kolb learning style inventory—version 3.1, technical specifications. 2005.

[55] A. Y. Kolb and D. Kolb. Learning styles and learning spaces: Enhancing experiential learning in higher education. *Academy of Management Learning & Education*, 4(2):193–212, 2005.

[56] D. A. Kolb. Individual learning styles and the learning proess. working paper #535-71. 1971.

[57] D. A. Kolb and R. Fry. Toward an applied theory of experiential learning. *C. Cooper (ed.), Theories of Group Process*, 1975.

[58] S. Kraemer. *Symbolische Maschinen: die Idee der Formalisierung im geschichtlichen Abriss*.

[59] T. Kuhn. *Die Struktur wissenschaftlicher Revolutionen. 24. Aufl*. Frankfurt am Main: Suhrkamp, 2007.

[60] R. Lehmann. *Lernstile als Grundlage adaptiver Lernsysteme in der Softwareschulung*. Munster [u.a.]: Waxmann, 2010.

[61] T. Leidig. L3–towards an open learning environment. *Journal on Educational Resources in Computing*, (1), 2001.

[62] N. Manouselis, H. Drachsler, R. Vuorikari, H. G. K. Humme, and R. Koper. Recommender systems in technology enhanced learning. pages 387–415, 2011.

[63] A. Martens. Adaptivität in hypermedialen lernsystemen. *Zeitschrift für eLearning*, 2008.

[64] M. McLuhan. *Understanding Media. The Extensions of Man.* McGraw-Hill, New York, 1964.

[65] N. Meder. Didactic requirements of learning environments: the web didactics approach of l3. *E-Learning Services in the Crossfire: Pedagogy, Economy, and Technology.*

[66] N. Meder. *Web-Didaktik. Eine neue Didaktik webbasierten, vernetzten Lernens.* Bertelsmann: Bielefeld, 2006.

[67] S. E. Metros. Learning objects in higher education. *Educause Research Bulletin*, 19:2–10, 2002.

[68] A. Mumford. Putting learning styles to work. *Action Learning at Work*, pages 121–135, 1997.

[69] R. Neches, T. Finin R. Fikes and, T. Gruber, R. Patil, T. Senator, and W. R. Swartout. Enabling technology for knowledge sharing. *AI Magazine*, 12:37–56, 1991.

[70] M. Neteler, M. H. Bowman, and M. Metz M. Landa. A multi-purpose open source gis. *Environmental Modelling & Software*, 31:124–130, 2012.

[71] A. Newell, J. C. Shaw, and H. A. Simon. Report on a general problem-solving program. In *Proceedings of the International Conference on Information Processing*, pages 256–264, 1959.

[72] H. S. Nwana. Intelligent tutoring systems: an overview. *Artificial Intelligence Review*, 4:251–277, 1990.

[73] H. S. Nwana. Intelligent tutoring systems: an overview. *Artificial Intelligence Review*, 4:251–277, 1990.

[74] J. Overhoff. *Die Frühgeschichte des Philanthropismus 1715–1771. Konstitutionsbedingungen, Praxisfelder und Wirkung eines pädagogischen Reformprogramms im Zeitalter der Aufklärung.* Tübingen: Niemeyer, 2004.

[75] M. Parmentier. Der bildungswert der dinge. *Zeitschrift für Erziehungswissenschaft*, 4(1):39–50, 2001.

[76] M. Pivec. Play and learn: potentials of game-based learning. *British Journal of Educational Technology*, 38(3):387–393, 2007.

[77] S. Pressey. A simple apparatus which gives tests and scores – and teaches. *School and Society*, 586:373–376, 1923.

[78] S. L. Pressey. A machine for automatic teaching of drill material. *School and Society*, (25):549–552, 1927.

[79] S. L. Pressey. A third and fourth contribution toward the coming industrial revolution in education. *School and Society*, (36):668–672, 1932.

[80] R. Reichenbach. Demokratisches selbst und dilettantisches subjekt. demokratische bildung und erziehung in der spätmoderne. 1999.

[81] L. Relly, H.-J. Schek, O. Henricsson, and S. Nebiker. Physical database design for raster images in concert. In *5th International Symposium on Spatial Databases (SSD'97)*, 1997.

[82] L. Relly, H. Schuldt, and H. Schek. Exporting database functionality – the concert way. *IEEE Data Eng. Bull 01/1998*, pages 43–51, 1998.

[83] P. Rigaux, M. Scholl, L. Segoufin, and S. Grumbach. Building a constraint-based spatial database system: model, languages, and implementation. *Inf. Syst. 28(6)*, pages 563–595, 2003.

[84] M. Rodrigues, S. Gonçalves, D. Carneiro, P. Novais, and F. Fdez-Riverola. Keystrokes and clicks: Measuring stress on e-learning students. In *Management Intelligent Systems: Second International Symposium*, pages 119–126, 2013.

[85] J. Ruhloff. *Das ungelöste Normproblem der Pädagogik. Eine Einführung*. Heidelberg: Verlag Quelle & Meyer, 1979.

[86] K. Schaaff, R. Degen, N. Adler, and M. T. P. Adam. Measuring affect using a standard mouse device. *Biomedical Engineering*, 57:761–764, 2012.

[87] Friedrich Schiller. *Über die ästhetische Erziehung des Menschen*. 1794.

[88] F. Schleiermacher. *Pädagogische Schriften*. Erich Weniger, unter Mitwirkung von Theodor Schulze, Düsseldorf: Schwann, 1957.

[89] D. Schmidt, M. Bleichenbacher, W. Dreyer, D. Heimberg, R. Italia, T. Mäder, T. Mauch, and C. Osterwalder. Calanda – a complete solution for time series management in banking, ubs, zurich.

[90] R. Schulmeister. *eLearning: Einsichten und Aussichten*. München: Oldenbourg, 2006.

[91] R. Schulmeister. *Grundlagen hypermedialer Lernsysteme. Theorie – Didaktik – Design*. Oldenbourg: München, 2007.

[92] B. E. Skinner. Teaching machines. *Science*, 128:969–977, 1958.

[93] B. E. Skinner. Programmed instruction revisited. *Phi Delta Kappan*, pages 103–110, 1986.

[94] H. Stachowiak. *Allgemeine Modelltheorie*. Wien, New York: Springer, 1973.

[95] A. Steiner. *A Generalisation Approach to Temporal Data Models and their Implementations*. dissertation, Swiss Federal Institute of Technology Zurich, 1998.

[96] Jun-Ming Su, Shian-Shyong Tseng, Jui-Feng Weng, Kuan-Ting Chen, Yi-Lin Liu, and Yi-Ta Tsai. An object based authoring tool for creating scorm compliant course. In *International Conference on Advanced Information Networking and Applications, IEEE*, volume 2, pages 950–951, 2002.

[97] Jun-Ming Su, Shian-Shyong Tseng, Jui-Feng Weng, Kuan-Ting Chen, Yi-Lin Liu, and Yi-Ta Tsai. An object based authoring tool for creating scorm compliant course. In *International Conference on Advanced Information Networking and Applications, IEEE*, volume 1, pages 209–214, 2005.

[98] C. Swertz. Computer als spielzeug. *Spektrum Freizeit*, 2:112–120, 1999.

[99] C. Swertz. überlegungen zur theoretischen grundlage der medien-pädagogik. pages 213–222, 2007.

[100] Y. Tang, L. Liang, R. Huang, and Y. Yu. Bitemporal extensions to non-temporal rdbms in distributed environments. In *Proceedings of the 8th International Conference on Computer Supported Cooperative Work in Design*, volume 2, pages 370–373, 2004.

[101] A. U. Tansel. Temporal relational data model. *IEEE Transactions on Knowledge and Data Engineering*, 9(3), 1997.

[102] Bo Kampman Walther. Playing and gaming. reflections and classifications. *Game Studies*, 3(1), 2003.

[103] R. Winter. *Die Kunst des Eigensinns. Cultural Studies als Kritik der Macht*. Weilerswist: Velbrück. Wittgenstein, L., 2001.

[104] L. Wittgenstein. *Tractatus logico-philosophicus, Logisch-philosophische Abhandlung*. Frankfurt am Main: Suhrkamp, 2003.

[105] J. Xu and R. H. Güting. A generic data model for moving objects. *GeoInformatica 17:1*, pages 125–172, 2013.

[106] P. Zimmermann, S. Guttormsen, B. Danuser, and P. Gomeza. Affective computing—a rationale for measuring mood with mouse and keyboard. *International Journal of Occupational Safety and Ergonomics*, 9(4):539–551, 2003.

Index

About the Editors

Kevin Fuchs holds a diploma in computer science. He has several years of professional experience in the operation computing centers, network monitoring and virtualization infrastructure. Currently he is doing active research on instructional design and learning analytics as part of his Ph.D. studies in education.

Prof. Dr. Peter A. Henning teaches computer graphics, semantic technologies, game programming and e-learning at Karlsruhe University of Applied Sciences since 1998 and Information Business Technology at the Steinbeis-University Berlin since 2012. He is founding director of the Institute of Computers in Education. He holds several degrees in theoretical physics (Diploma 1983, Doctorate 1987, Habilitation 1993) and has published more than 130 scientific articles and books. Prof. Henning is in charge of several industrial and scientific projects connected to technology enhanced learning, acts as the scientific committee director of the LEARNTEC, member of the program committee and evaluation board of the Virtual University of Bavaria and member of the board of the eLearning group within the GI (Gesellschaft fuer Informatik). Prof. Henning has initiated and is currently coordinating the FP7 EU project INTUITEL (Intelligent Tutorial Interfaces for Technology Enhanced Learning). Current research activities include the determination of the cognitive position of a learner in a multidimensional space of learning objects and its attribution to predefined learning pathways.